DEVELOPING SPATIAL THINKING

DEVELOPING SPATIAL THINKING

Sheryl Sorby

DELMAR
CENGAGE Learning

Australia • Brazil • Japan • Korea • Mexico • Singapore • Spain • United Kingdom • United States

DELMAR
CENGAGE Learning™

Developing Spatial Thinking
Workbook by Sheryl Sorby

Vice President, Editorial: Dave Garza

Director of Learning Solutions: Sandy Clark

Associate Acquisitions Editor: Kathryn Hall

Managing Editor: Larry Main

Product Manager: Mary Clyne

Editorial Assistant: Cris Savino

Software Development: Graphic World

Vice President, Marketing: Jennifer Baker

Marketing Director: Deborah Yarnell

Marketing Manager: Jillian Borden

Production Director: Wendy Troeger

Production Manager: Mark Bernard

Senior Content Project Manager: Mike Tubbert

Senior Art Director: Casey Kirchmayer

Technology Project Manager: Joe Pliss

For product information and technology assistance, contact us at
Cengage Learning Customer & Sales Support, 1-800-354-9706
For permission to use material from this text or product,
submit all requests online at **www.cengage.com/permissions**
Further permissions questions can be emailed to
permissionrequest@cengage.com

ISBN-13: 978-1-111-13906-3

ISBN-10: 1-111-13906-7

Delmar
Executive Woods
5 Maxwell Drive
Clifton Park, NY 12065
USA

Cengage Learning is a leading provider of customized learning solutions with office locations around the globe, including Singapore, the United Kingdom, Australia, Mexico, Brazil, and Japan. Locate your local office at **www.cengage.com/global**

Cengage Learning products are represented in Canada by Nelson Education, Ltd.

To learn more about Delmar, visit **www.cengage.com/delmar**

Purchase any of our products at your local bookstore or at our preferred online store **www.cengagebrain.com**

Notice to the Reader
Publisher does not warrant or guarantee any of the products described herein or perform any independent analysis in connection with any of the product information contained herein. Publisher does not assume, and expressly disclaims, any obligation to obtain and include information other than that provided to it by the manufacturer. The reader is expressly warned to consider and adopt all safety precautions that might be indicated by the activities described herein and to avoid all potential hazards. By following the instructions contained herein, the reader willingly assumes all risks in connection with such instructions. The publisher makes no representations or warranties of any kind, including but not limited to, the warranties of fitness for particular purpose or merchantability, nor are any such representations implied with respect to the material set forth herein, and the publisher takes no responsibility with respect to such material. The publisher shall not be liable for any special, consequential, or exemplary damages resulting, in whole or part, from the readers' use of, or reliance upon, this material.

Printed in the United States of America
4 5 6 7 18 17 16 15 14

CONTENTS

PREFACE

The ability to visualize in three dimensions has been shown to be an important skill for people who intend to study in scientific and technical fields. Well-developed spatial skills have been linked to success in engineering, computer science, chemistry, medicine, mathematics, and architecture to name just a few. Design is central to engineering and well developed spatial skills have been shown to be critical to a person's ability to develop creative design solutions to problems. Well developed spatial skills have also been linked to a person's ability to interact with a computer in performing database manipulations and to a person's ability to understand various aspects of structural chemistry. Doctors, who must learn to use modern-day laparoscopy tools, require well developed 3-D spatial skills. Architects must often visualize how a new structure will look as well as how it interacts with its surroundings when designing a new building.

In educational psychology research, the distinction is often made between "spatial ability" and "spatial skills." The difference between the two is described briefly in the following. Spatial ability is defined as the innate ability to visualize that a person has before any formal training has occurred; i.e., a person is born with ability. However, spatial skills are learned or are acquired through training. As with any other type of skill (writing, mathematics, etc.), some people might have a higher degree of innate ability than others; however, most people can eventually acquire the skill through patience and practice. The materials in this text will help you develop your 3-D spatial visualization skills.

These materials contain ten separate modules to help you develop your 3-D spatial visualization skills. For each module, there is a software component in addition to this workbook component. To maximize your skill development, we suggest that you first work through the appropriate software module. After you complete the software module, work through the pages for that module in this workbook. Because sketching with paper and pencil have been shown to be particularly helpful in the development of 3-D spatial skills, you will find considerable benefit in completing the workbook pages, many of which require hand sketching. You should probably work through the modules in the order in which they are presented in the software and in this workbook, but you can do them in any order you like. One exception to this is that you should complete and understand Module 7 (Rotation of Objects about a Single Axis) before you attempt Module 8 (Rotation of Objects about Two or More Axes) because these two modules are meant to be completed sequentially.

The multimedia software is available on CD-ROM or on-line at www.cengagebrain.com. The software works on either a PC or Macintosh platform and requires no additional software to run.

Good luck and have fun!

ACKNOWLEDGMENTS

The author and Publisher wish to acknowledge several individuals for their contributions to this project:

- For the original vision: Beverly Baartmans
- For the original worksheets: David Shaffer
- For Workbook revisions: Jeff Valensky
- For software updates: Judith Birchman

This material is based upon work supported by the National Science Foundation under Grant No. DUE-9752660. Any opinions, findings, and conclusions or recommendations expressed in this material are those of the authors and do not necessarily reflect the views of the National Science Foundation.

Surfaces and Solids of Revolution

A surface of revolution is like a hollow shell created by revolving a set of 2-D curves about a coordinate axis or about another line in 3-D space.

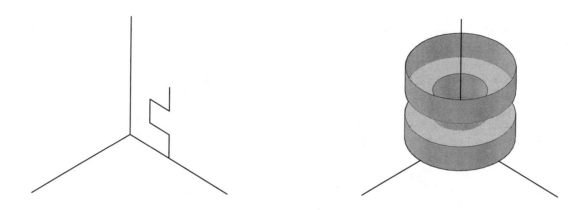

2-D Shape and Surface of Revolution

A solid of revolution is a 3-D object of a finite volume. It is generated by revolving a closed 2-D shape about a coordinate axis of another line in 3-D space.

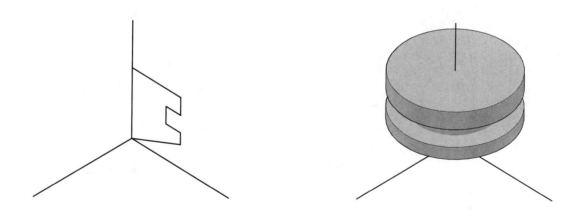

Closed 2-D Shape and Solid of Revolution

The resulting solid or surface of revolution will vary depending on the chosen axis of revolution.

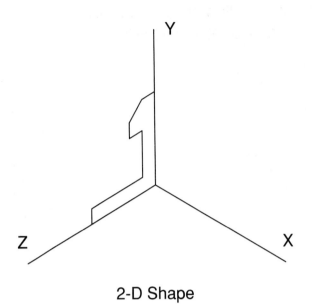

2-D Shape

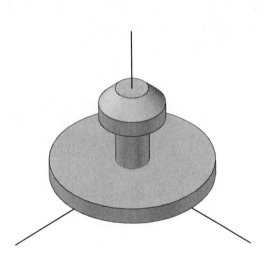

Shape Revolved About Y

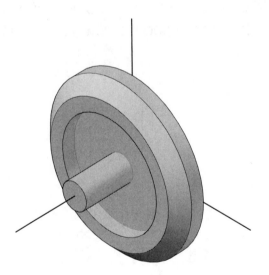

Shape Revolved About Z

When 2-D shapes are revolved about an axis, you can choose to revolve them less than 360 degrees with different results.

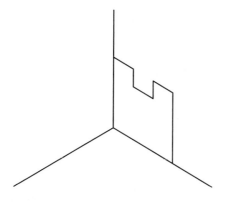

2-D Shape

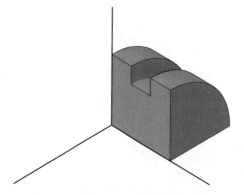

Shape Revolved 90 Degrees

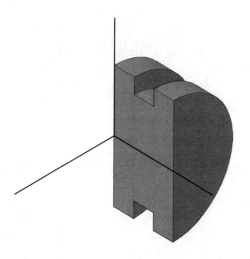

Shape Revolved 180 Degrees

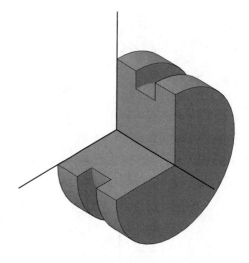

Shape Revolved 270 Degrees

Different objects will be formed depending on a combination of the choice for the axis of revolution and the choice of the angle of revolution.

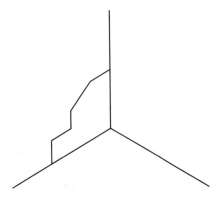

2-D Shape

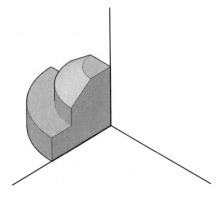

Shape Revolved 90 Degrees About Y

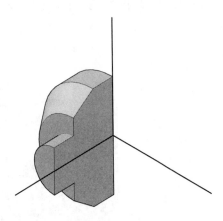

Shape Revolved 180 Degrees About Z

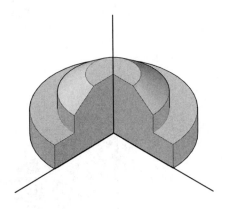

Shape Revolved 270 Degrees About Y

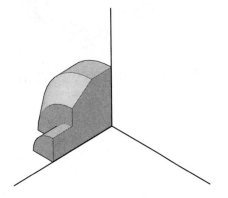

Shape Revolved 90 Degrees About Z

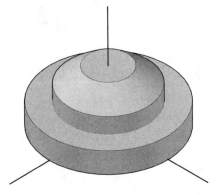

Shape Revolved 360 Degrees About Y

REV - IV

If the axis of revolution is not located on the 2-D shape itself but some distance, x, away from it, then a solid of revolution with a cylindrical hole of diameter 2x will be created from it.

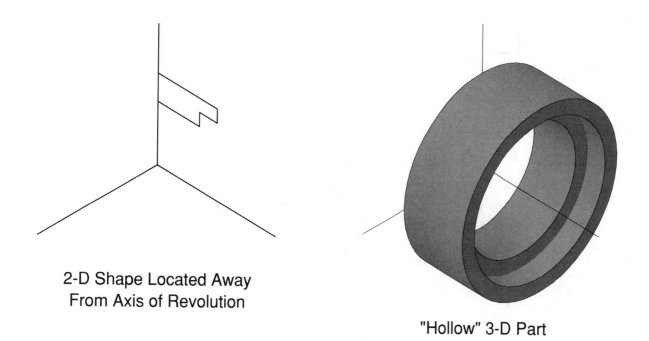

2-D Shape Located Away
From Axis of Revolution

"Hollow" 3-D Part

To visualize what a shape will look like after it has been revolved around an axis, first think about mirroring the shape on the opposite side of the axis of revolution. Then form an object with a basic cylindrical outline through the two shapes.

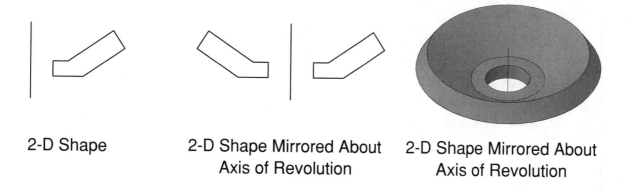

2-D Shape

2-D Shape Mirrored About
Axis of Revolution

2-D Shape Mirrored About
Axis of Revolution

Circle the letter corresponding to the object or objects that were formed by revolving the shape shown on the left about an axis. There may be more than one answer per problem.

1.

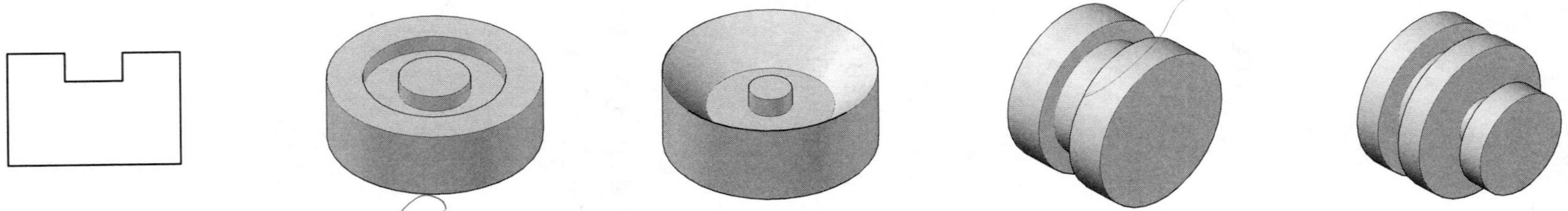

A   B   C   D

2.

A   B   C   D

3.

A   B   C   D

Name: Section: Date:

Class:

Developing Spatial Thinking

Grade: Page
                         rev—01

Circle the letter corresponding to the object or objects that were formed by revolving the shape shown on the left about an axis. There may be more than one answer per problem.

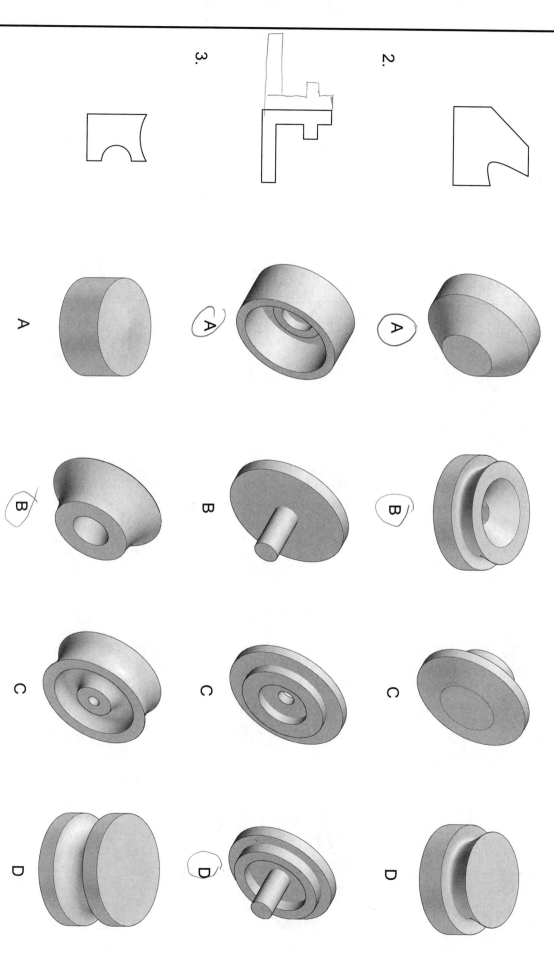

1.

A (circled)

B

C

D

2.

A

B

C

D (circled)

3.

A

B (circled)

C

D

Name:
Class:

Section:

Date:

Developing Spatial Thinking

Grade:

Page
rev—02

Circle the letter corresponding to the object or objects that were formed by revolving the shape shown on the left about an axis. There may be more than one answer per problem.

1.

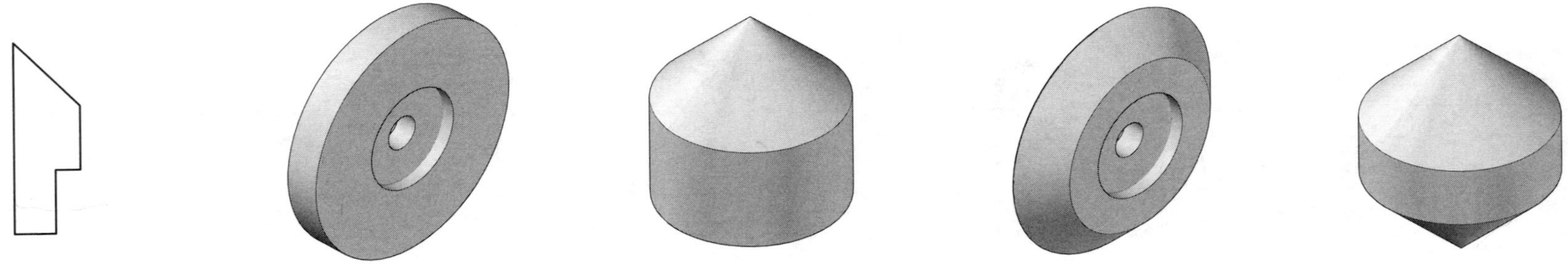

A ✗ B ✗ C D ✗

2.

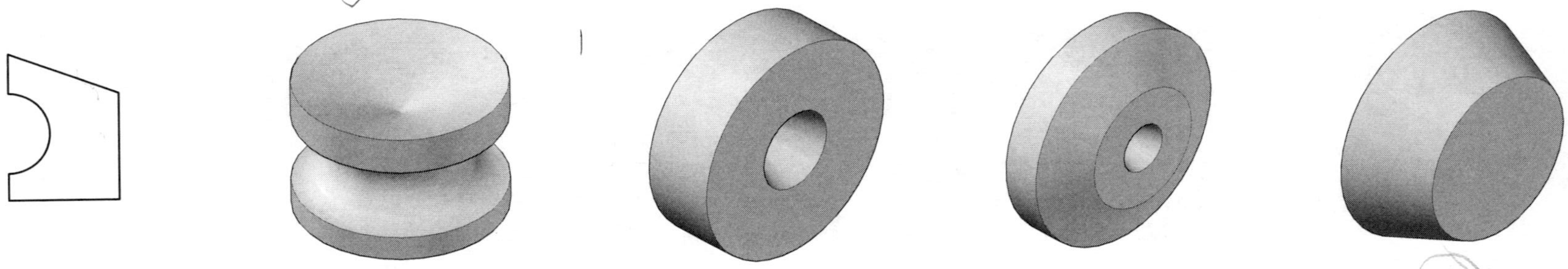

A B C D

3.

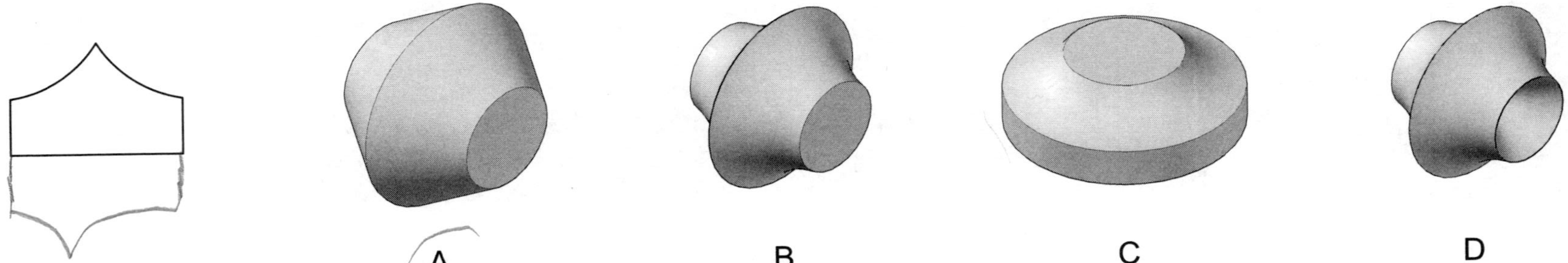

(A) B C D

Circle the letter corresponding to the object or objects that were formed by revolving the shape shown on the left about an axis. There may be more than one answer per problem.

1.

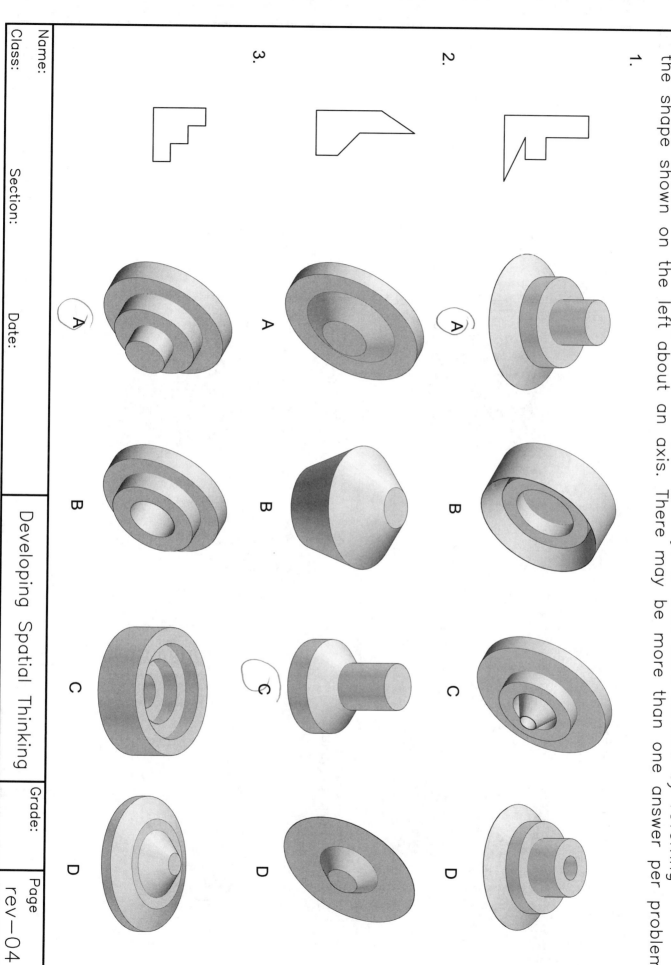

2.

3.

Name:

Class:

Section:

Date:

Developing Spatial Thinking

Grade:

Page
rev—04

Indicate the axis about which the 2-D shape was revolved to obtain the given solid.

1.

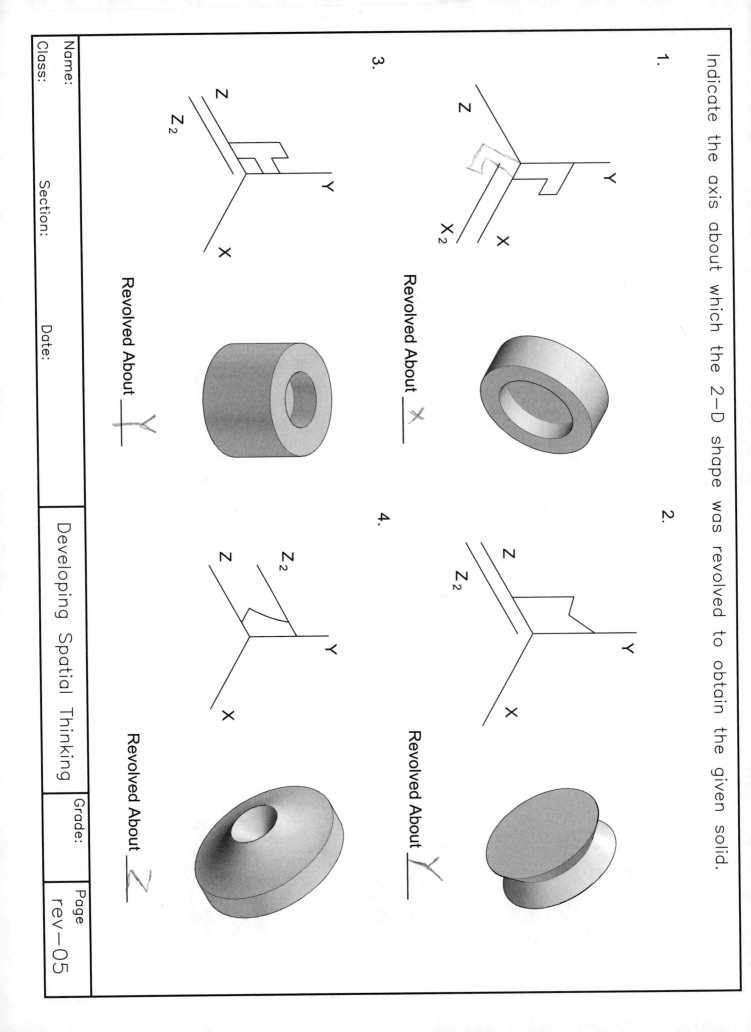

Revolved About __X__

2.

Revolved About ____

3.

Revolved About __Y__

4.

Revolved About __Z__

Indicate the axis about which the 2-D shape was revolved to obtain the given solid.

1.

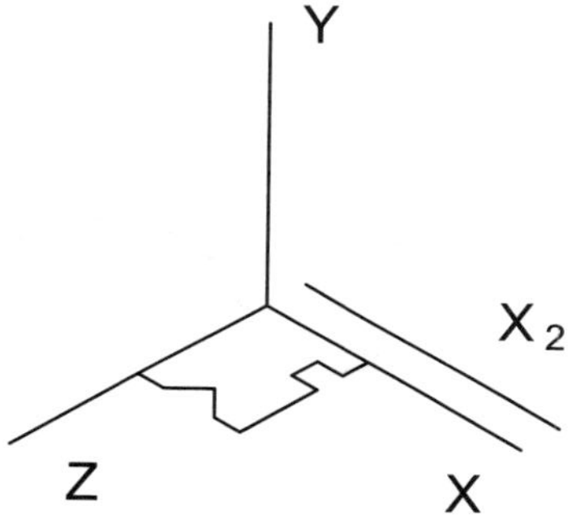

Revolved About ___X___

2.

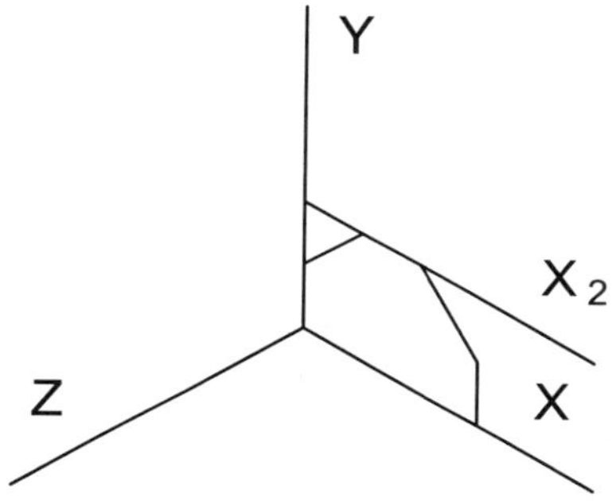

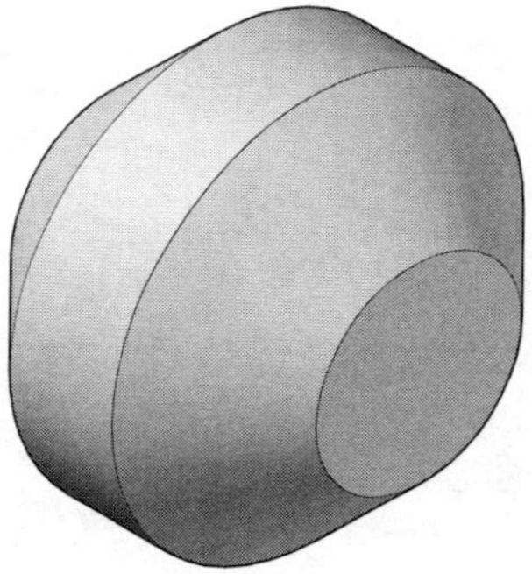

Revolved About ___X₂___

3.

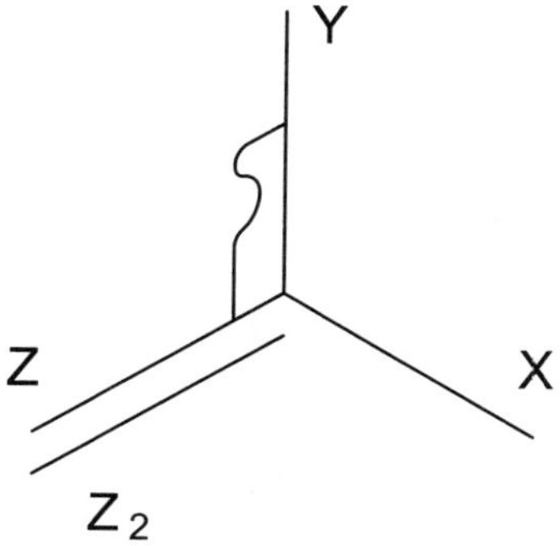

Revolved About ___Z___

4.

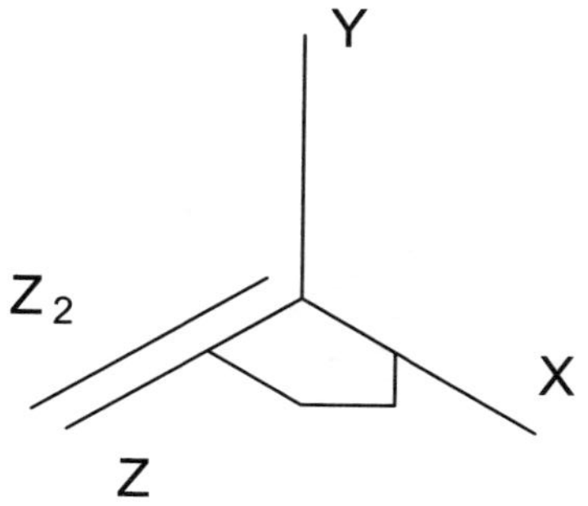

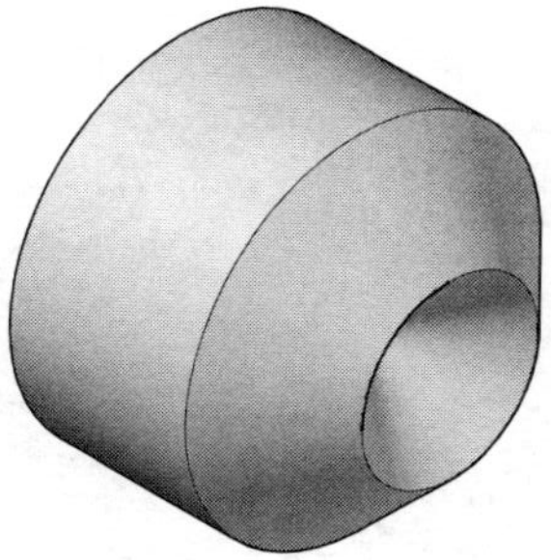

Revolved About ___X___

Name:

Class:

Section:

Date:

Developing Spatial Thinking

Grade:

Page
rev-06

Indicate the axis about which the 2-D shape was revolved to obtain the given solid.

1.

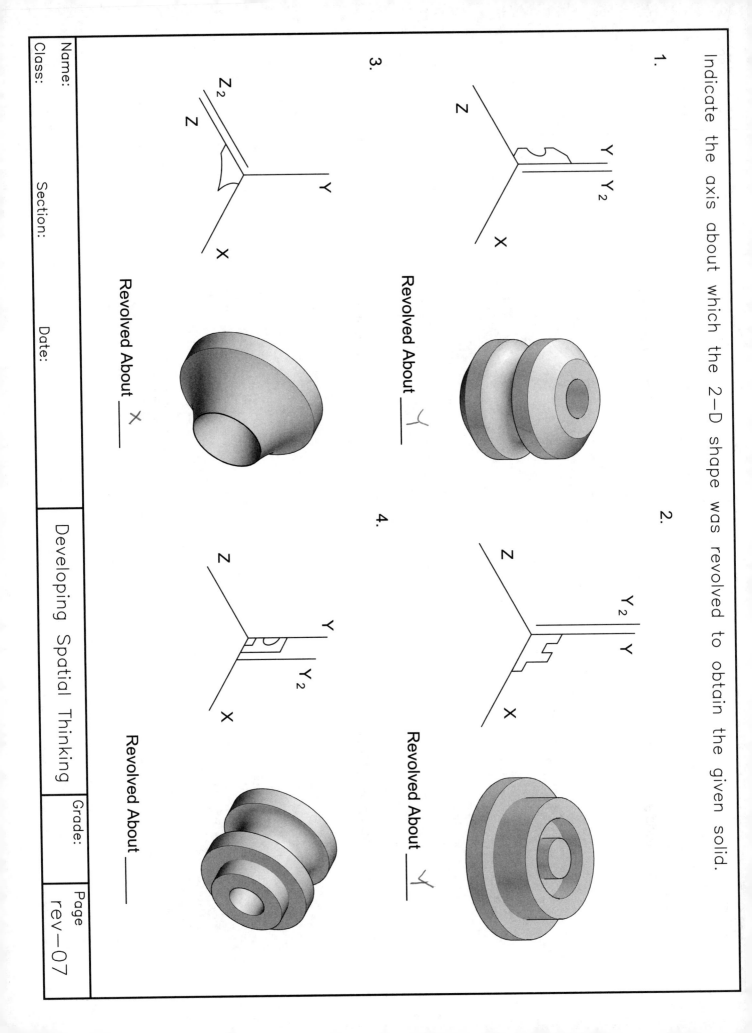

Revolved About ___ Y

2.

Revolved About ___ Y

3.

Revolved About ___ X

4.

Revolved About ___

Indicate the axis about which the 2–D shape was revolved to obtain the given solid.

1.

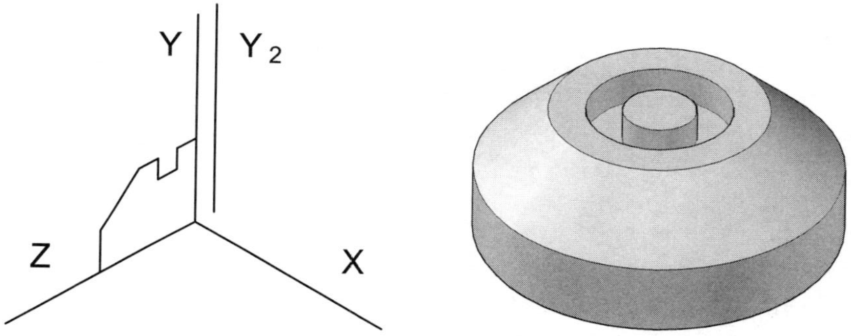

Revolved About Y

2.

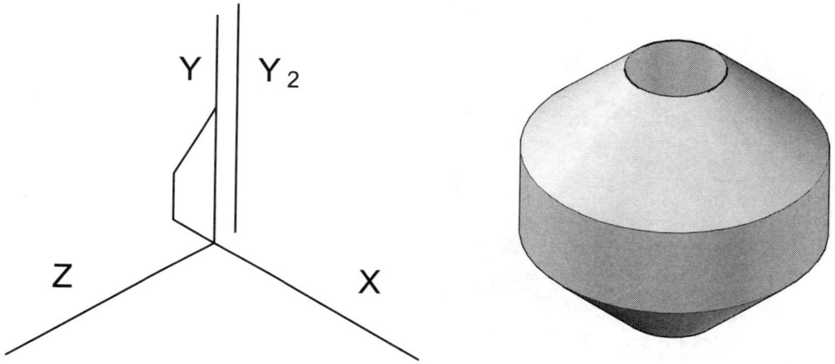

Revolved About Y

3.

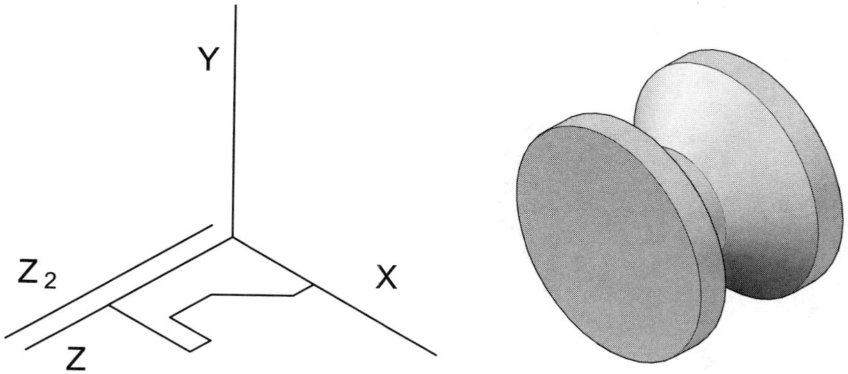

Revolved About Z

4.

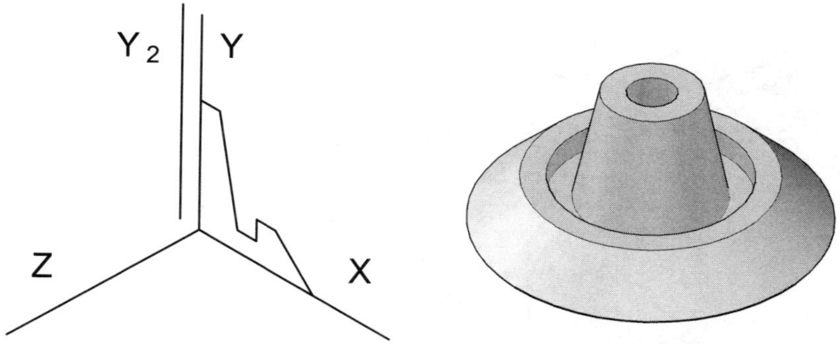

Revolved About Y

Name: Class: Section: Date:	Developing Spatial Thinking	Grade:	Page rev–08

Indicate the axis and the number of degrees (90, 180, 270, or 360) about which the 2-D shape was revolved to obtain the given solid.

1.

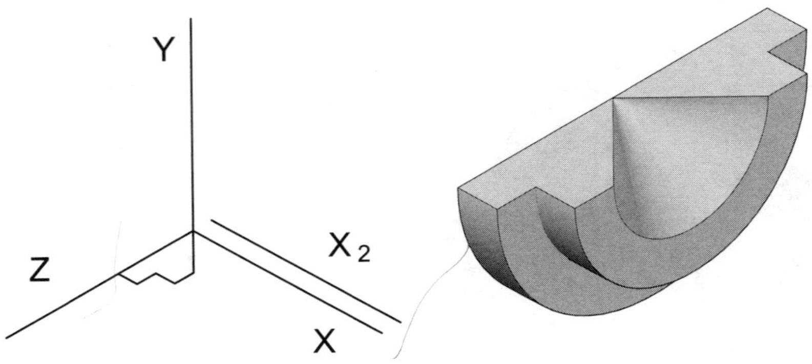

Z

Y

X₂

X

Revolved ___Z___ About __180°__

2.

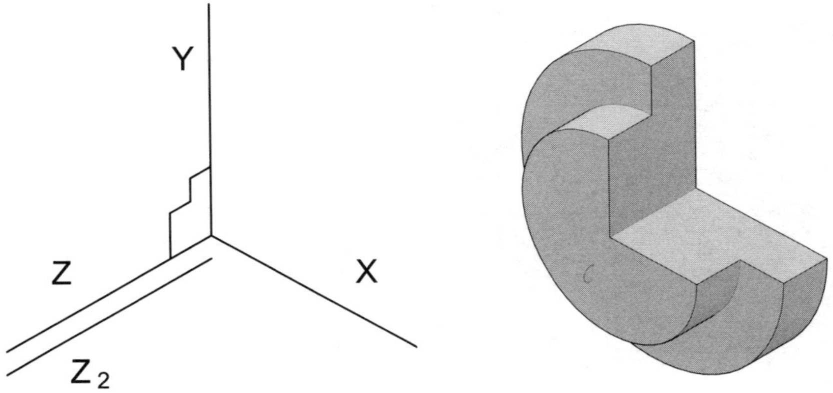

Z

Z₂

Y

X

Revolved ___Z___ About __270°__

3.

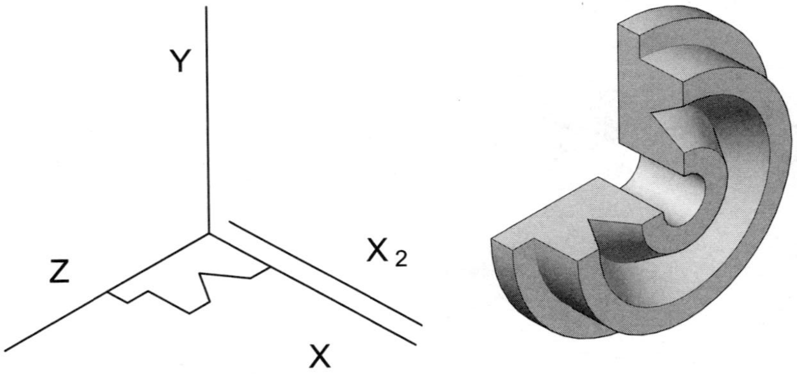

Z

Y

X

X₂

Revolved ___X___ About __270°__

4.

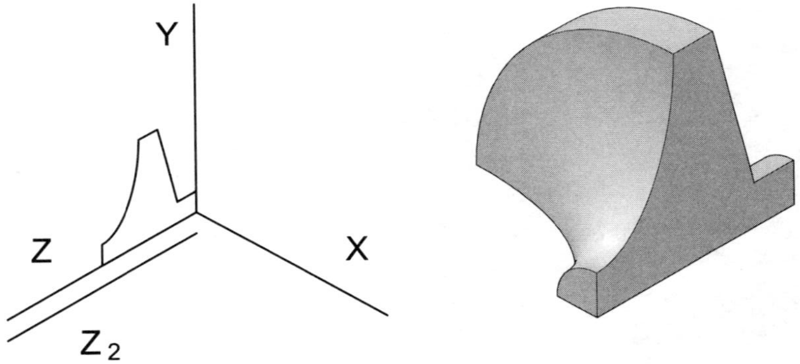

Z

Z₂

Y

X

Revolved ___Y___ About __90°__

Indicate the axis and the number of degrees (90, 180, 270, or 360) about which the 2-D shape was revolved to obtain the given solid.

1.

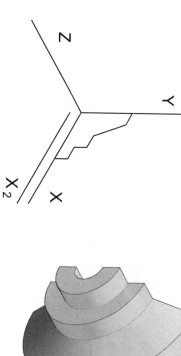

Revolved __Y__ About __180°__

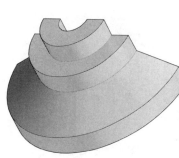

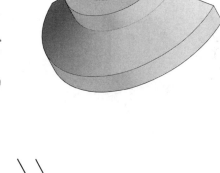

2.

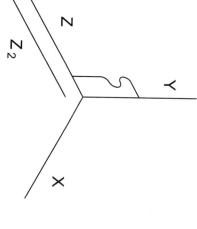

Revolved __Z__ About __270°__

3.

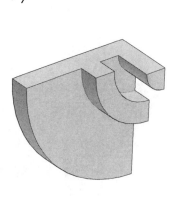

Revolved __Z__ About __90°__

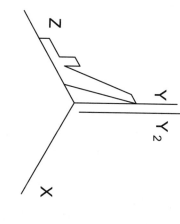

4.

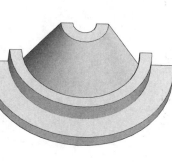

Revolved __Z__ About __180°__

Name:
Class:

Section:

Date:

Developing Spatial Thinking

Grade:

Page
rev—10

Indicate the axis and the number of degrees (90, 180, 270, or 360) about which the 2-D shape was revolved to obtain the given solid.

1.

Revolved _____ About _____

2.

Revolved _X_ About _90°_

3.

Revolved _Z_ About _90_°

4.

Revolved _X_ About _180_°

Name: Section:
Class: Date:

| Developing Spatial Thinking | Grade: | Page |
| | | rev-11 |

Indicate the axis and the number of degrees (90, 180, 270, or 360) about which the 2-D shape was revolved to obtain the given solid.

1.

Z

Y   Y₂

X

Revolved __Z__ About __90°__

2.

Z

Y₂   Y

X

Revolved __X__ About __270°__

3.

Z

Y

X₂   X

Revolved __Y__ About __180°__

4.

Z

Y

X₂   X

Revolved __Z__ About __90°__

Name: Section:
Class: Date:

Developing Spatial Thinking

Grade:

Page
rev-12

For the objects shown on the left below, circle the letter corresponding to the shape that was revolved to create it.

1.

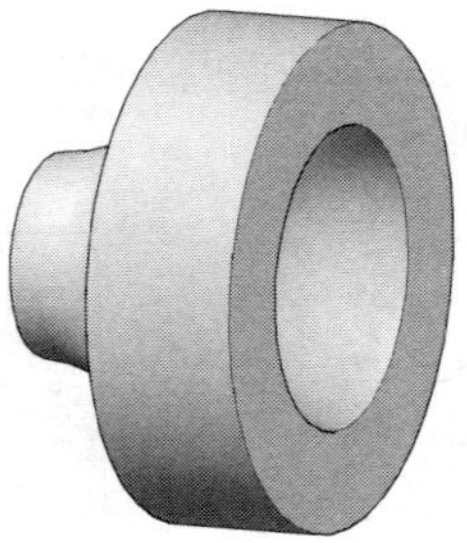

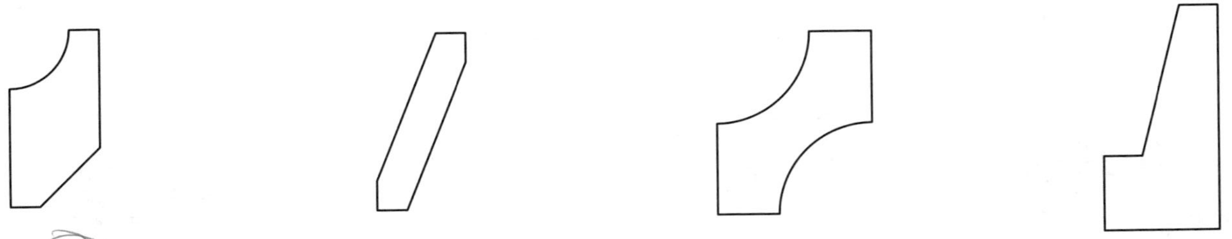

A

B

C

D

2.

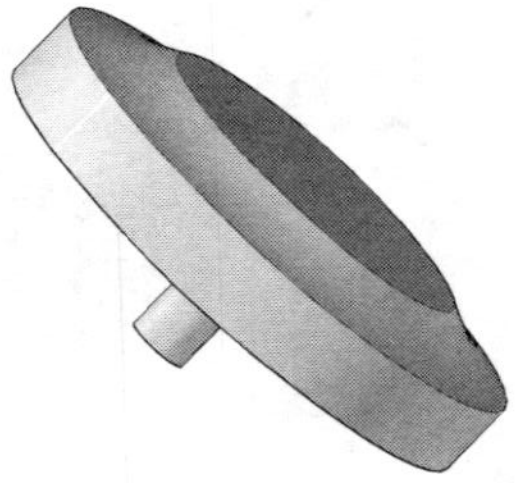

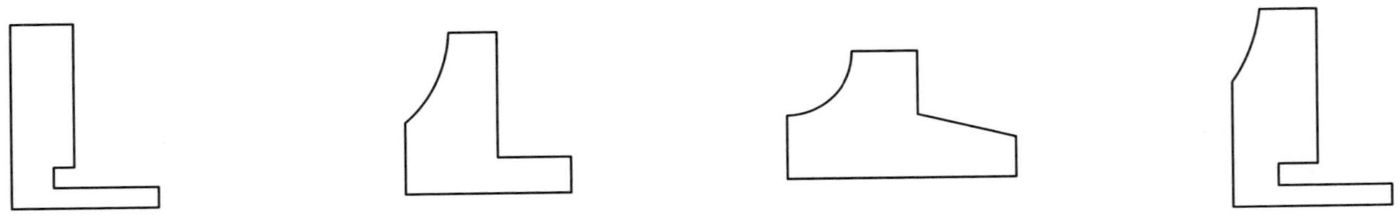

A

B

C

D

3.

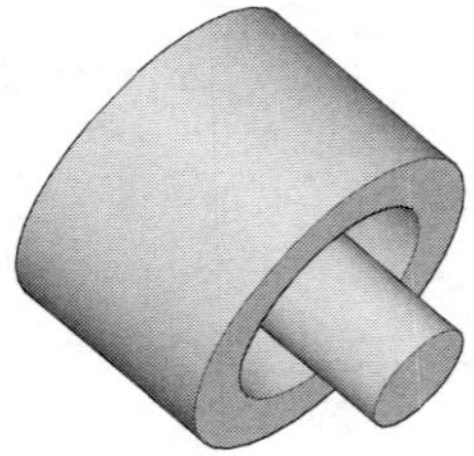

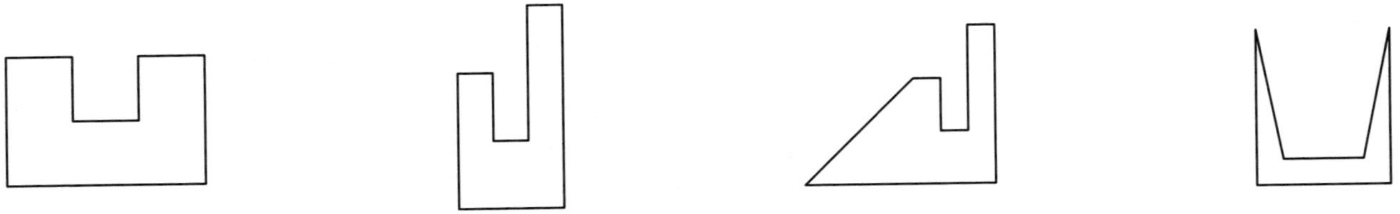

A

B

C

D

For the objects shown on the left below, circle the letter corresponding to the shape that was revolved to create it.

1.

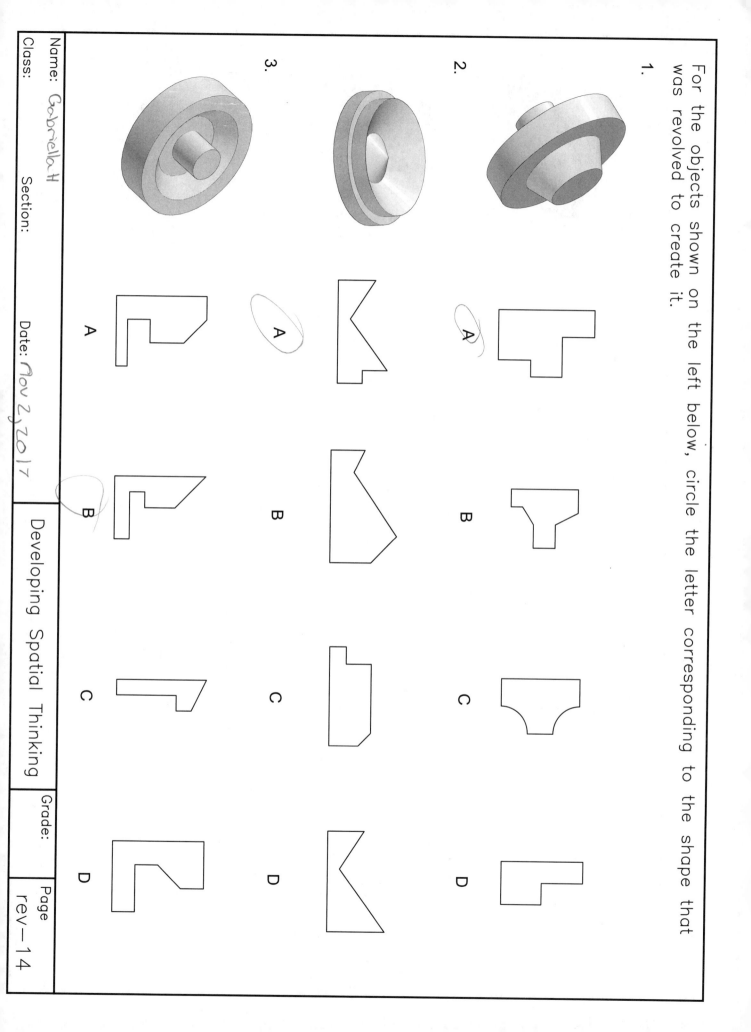

A

B

C

D

2.

A

B

C

D

3.

A

B

C

D

Name: Gabriella H

Class:

Section:

Date: Nov 2, 2017

Developing Spatial Thinking

Grade:

For the objects shown on the left below, circle the letter corresponding to the shape that was revolved to create it.

1.

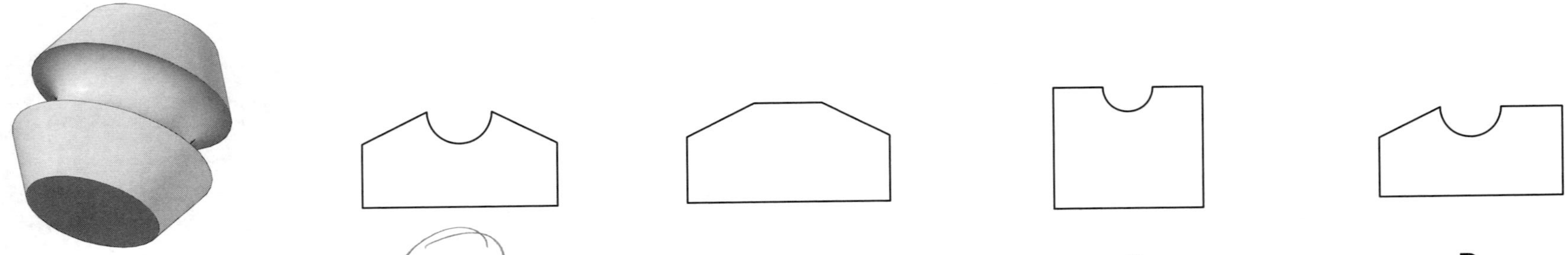

A B C D

2.

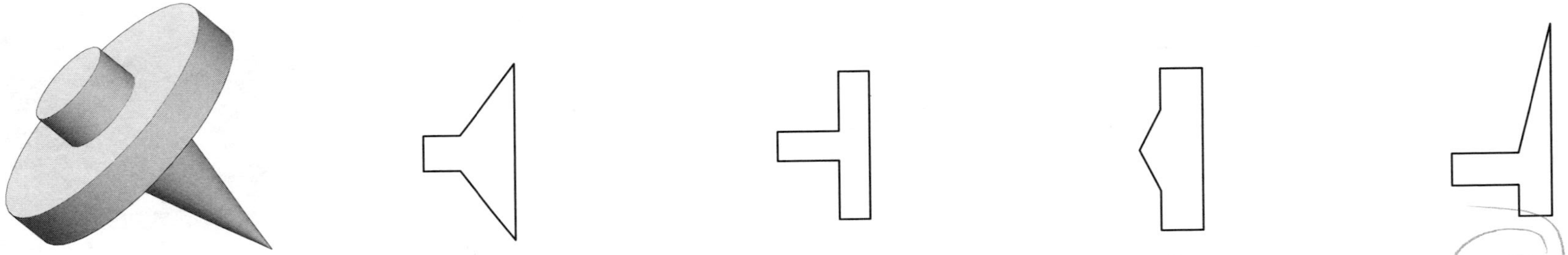

A B C D

3.

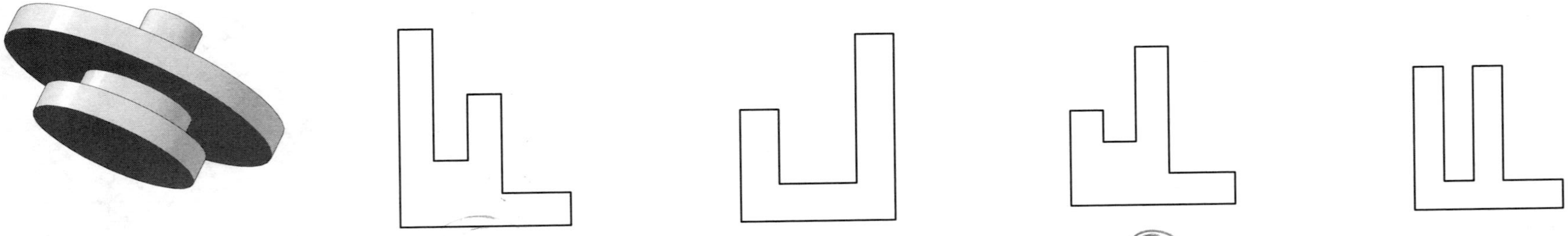

A B C D

Name: Class: Section: Date:	Developing Spatial Thinking	Grade:	Page rev–15

Combining Solid Objects

Two overlapping objects can be combined to make a new object by joining, cutting, or intersecting.

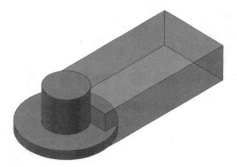

Original Two Objects

Objects Joined

Objects Cut

Objects Intersected

The volume of interference is the volume that two overlapping objects have in common. The different combining operations--joining, cutting, and intersecting, use the volume of interference in different ways.

Two Overlapping Objects

Volume of Interference

When two objects are joined, they become a single object. The volume of interference is absorbed into the resulting object.

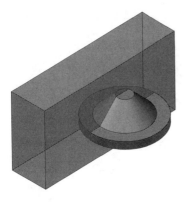

Two Overlapping Objects

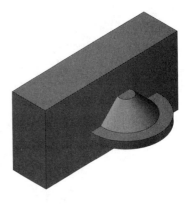

Objects Joined

When two objects are combined by cutting, the volume of interference is removed from the object being cut.

Two Overlapping Objects

Objects Cut

In a cutting operation, one object acts as a cutting tool on the other object. The final result depends on which object is designated as the cutting tool and which object is the object being cut.

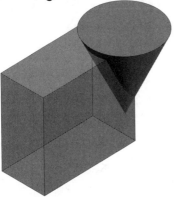

Overlapping Objects

Cone Cuts Block

Block Cuts Cone

When an intersect operation is performed, the resulting object consists of the volume that is common to the two original objects (the volume of interference).

Two Overlapping Objects

Intersected Objects

More complicated objects can be created by a series of cut/join/intersect operations. In this situation, it is best to look at the final object and think about the building blocks that were part of its creation. The creation of a more complex part is illustrated in the following figures.

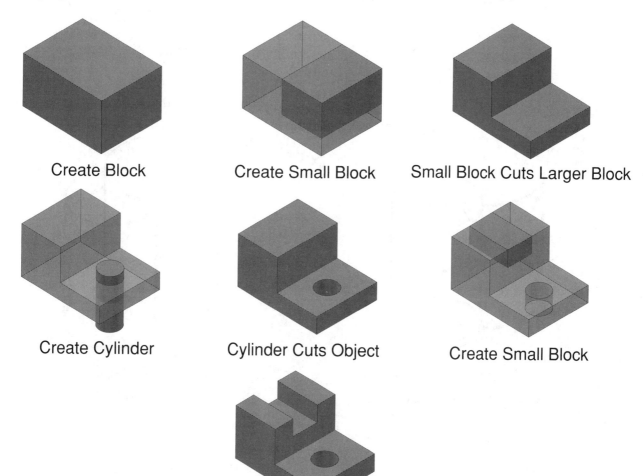

Create Block   Create Small Block   Small Block Cuts Larger Block

Create Cylinder   Cylinder Cuts Object   Create Small Block

Small Block Cuts Object -- Desired Result

Sometimes you will be presented with two overlapping objects and must try to visualize the result of combining the objects. If you think about the edges and surfaces that define the object and then try to imagine which of them will be part of the final object and which will be removed, you can begin to fill in the shape of the result of the combining operation. In the following figure, two overlapping cylinders are shown. Also shown are the results of a join, an intersect, and two cut operations (the larger cylinder cutting the smaller one and vise versa).

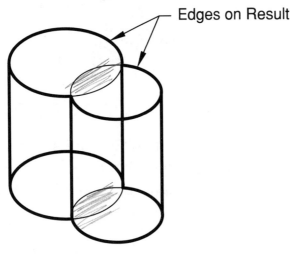

Cylinders Joined

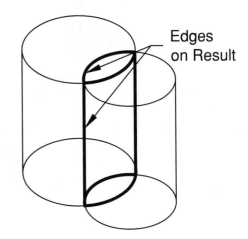

Cylinders Intersected

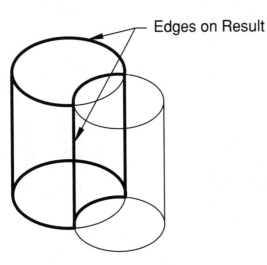

Small Cylinder
Cuts Large Cylinder

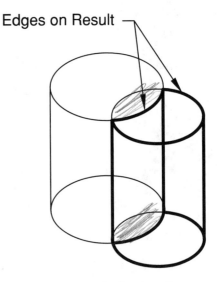

Large Cylinder Cuts
Small Cylinder

The objects on the left are to be combined, with the result shown on the right. Circle the appropriate word (cut, join, or intersect) indicating the operation that was performed.

1.

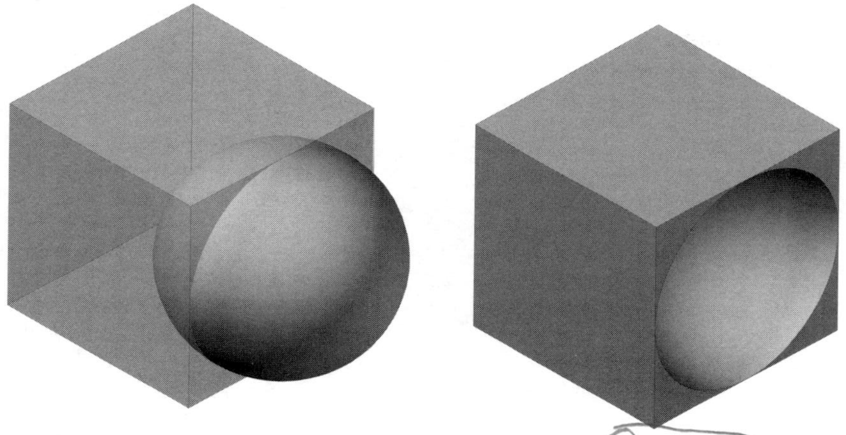

Operations Performed: Cut Join Intersect

2.

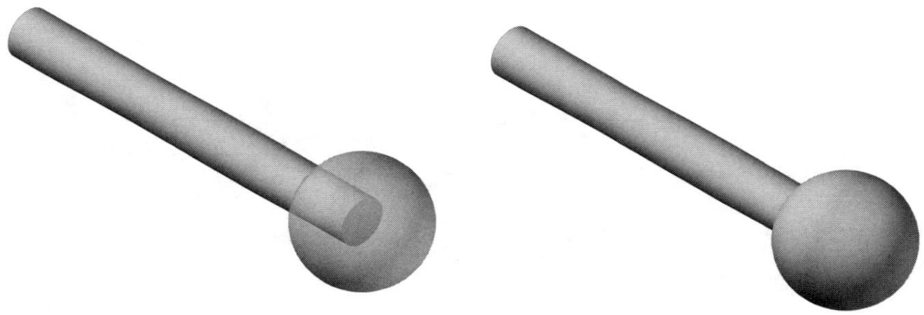

Operations Performed: Cut Join Intersect

3.

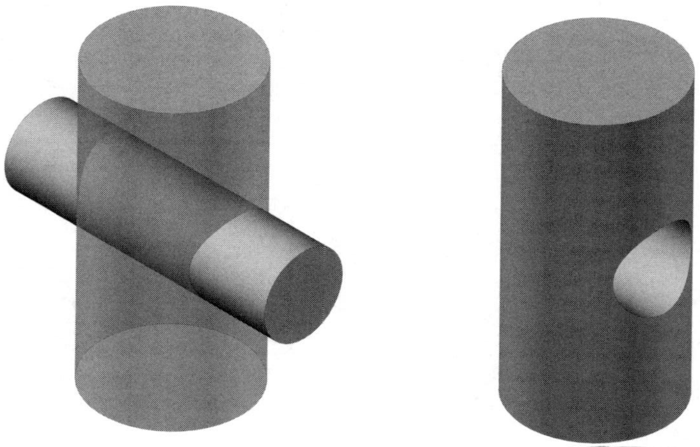

Operations Performed: Cut Join Intersect

4.

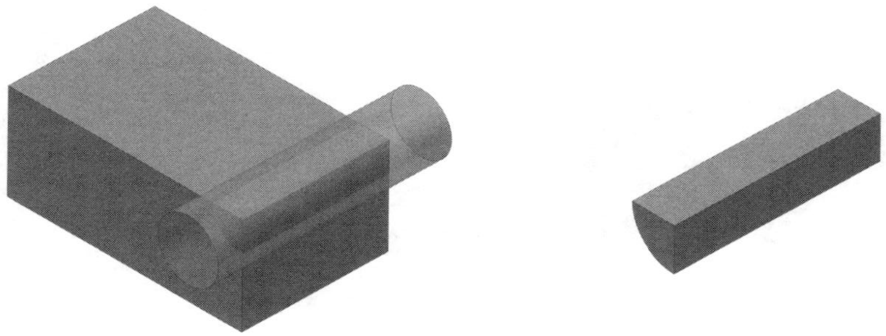

Operations Performed: Cut Join Intersect

Name: Class: Section: Date:	Developing Spatial Thinking	Grade:	Page ob/int-01

The objects on the left are to be combined, with the result shown on the right. Circle the appropriate word (cut, join, or intersect) indicating the operation that was performed.

1.

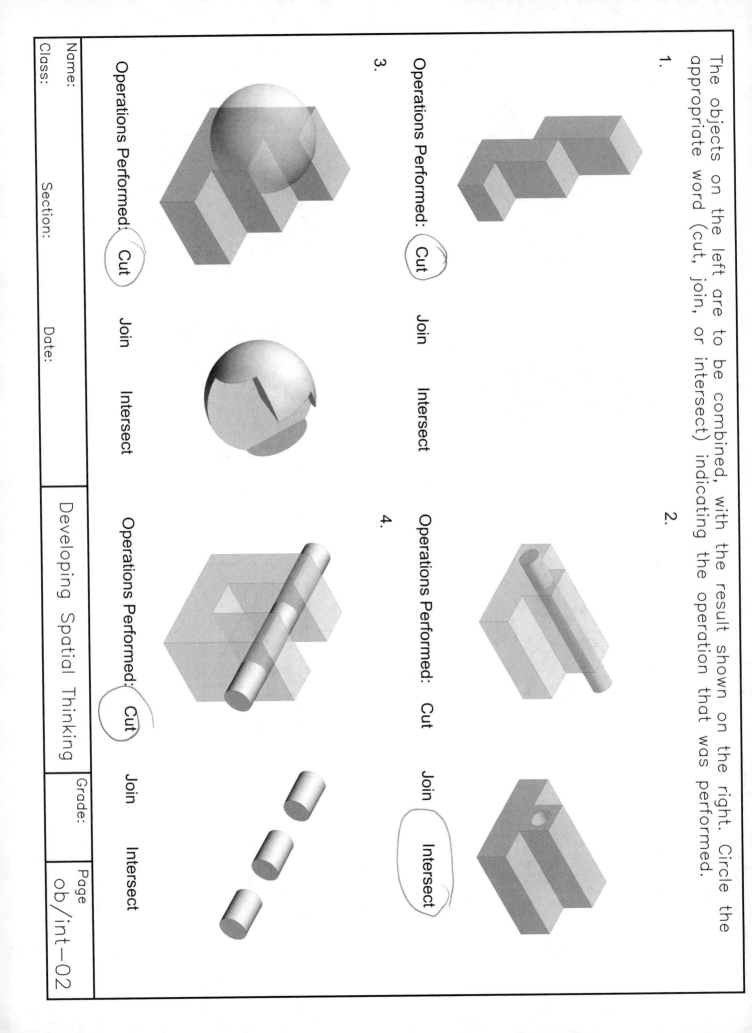

Operations Performed: (Cut)   Join   Intersect

2.

Operations Performed: Cut   Join   (Intersect)

3.

Operations Performed: (Cut)   Join   Intersect

4.

Operations Performed: (Cut)   Join   Intersect

The objects on the left are to be combined, with the result shown on the right. Circle the appropriate word (cut, join, or intersect) indicating the operation that was performed.

1.

Operations Performed: Cut   ⟨Join⟩   Intersect

2.

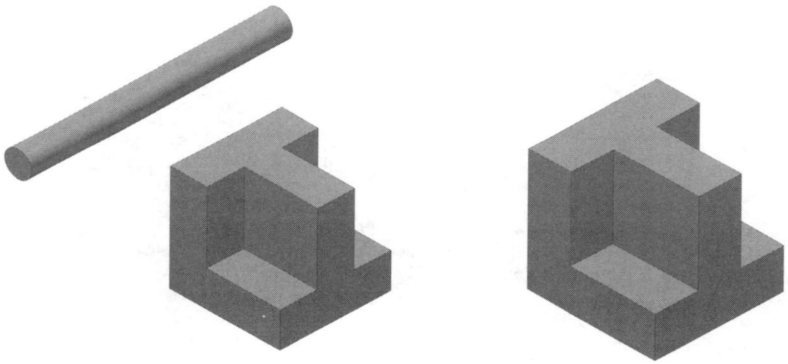

Operations Performed: Cut   ⟨Join⟩   Intersect

3.

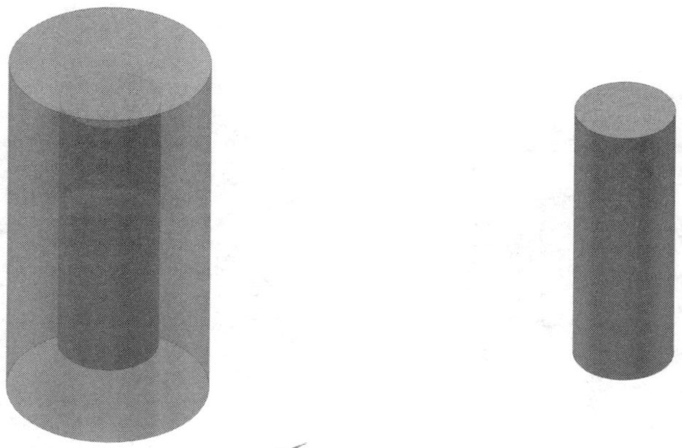

Operations Performed: ⟨Cut⟩   Join   Intersect

4.

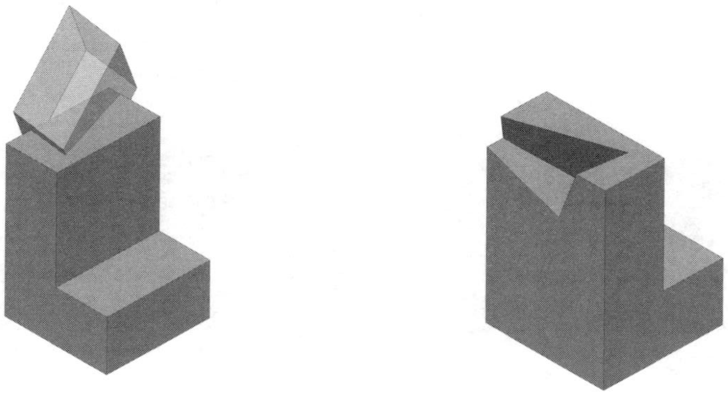

Operations Performed: ⟨Cut⟩   Join   Intersect

The objects on the left are to be combined, with the result shown on the right. Circle the appropriate word (cut, join, or intersect) indicating the operation that was performed.

1.

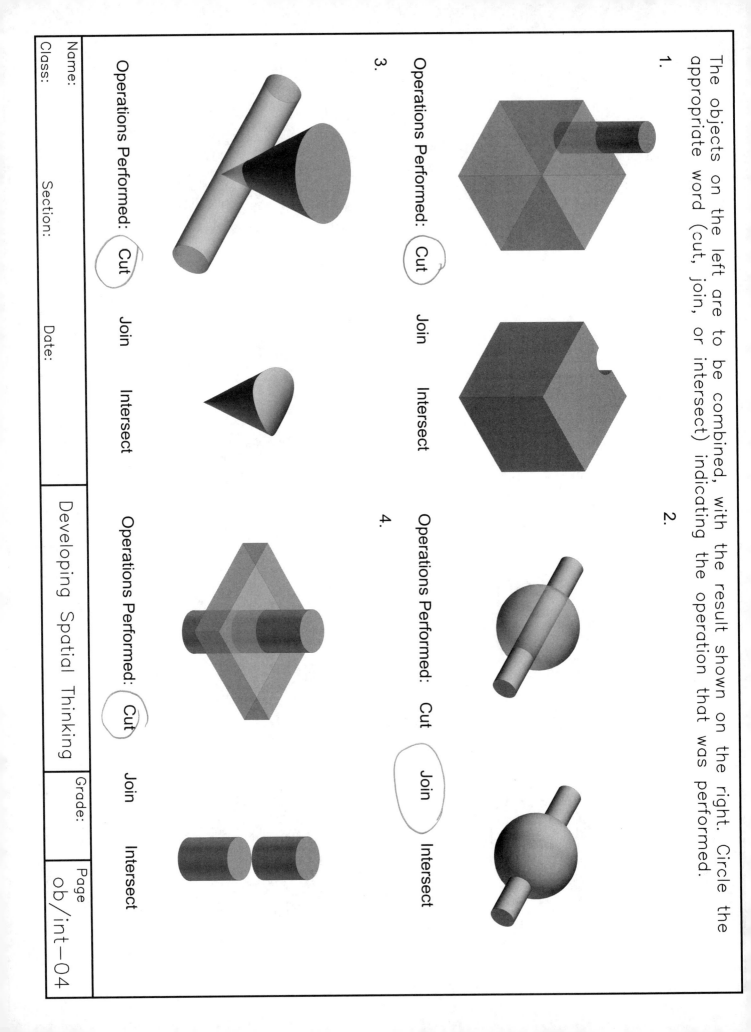

2.

Operations Performed: (Cut)   Join   Intersect

3.

Operations Performed: (Cut)   Join   Intersect

4.

Operations Performed: Cut   (Join)   Intersect

Operations Performed: (Cut)   Join   Intersect

Name:

Class:

Section:

Date:

Developing Spatial Thinking

Grade:

Page
ob/int—04

For the two overlapping objects shown on the left below, match the letter corresponding to the combining operation that was performed to obtain each of the three objects shown on the right.

A. Intersect   C. Object 1 cuts Object 2
B. Join   D. Object 2 cuts Object 1

1.

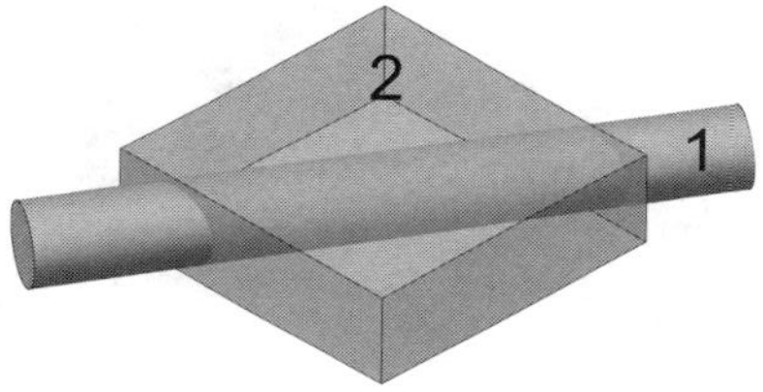

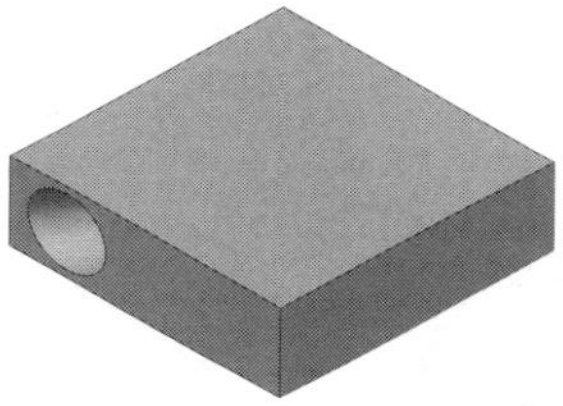

A

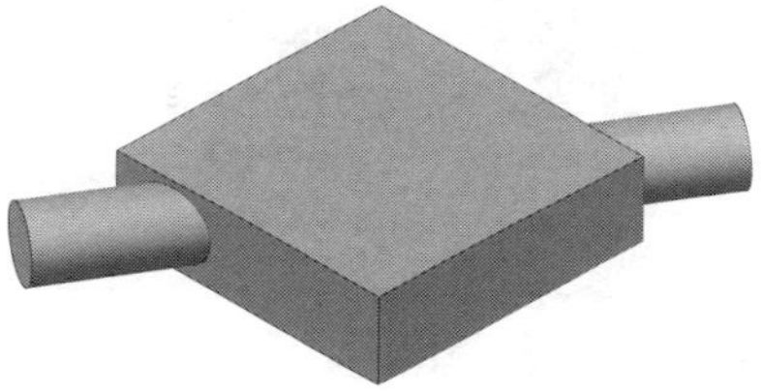

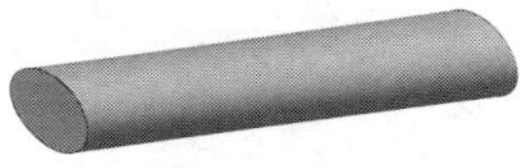

C

B

2.

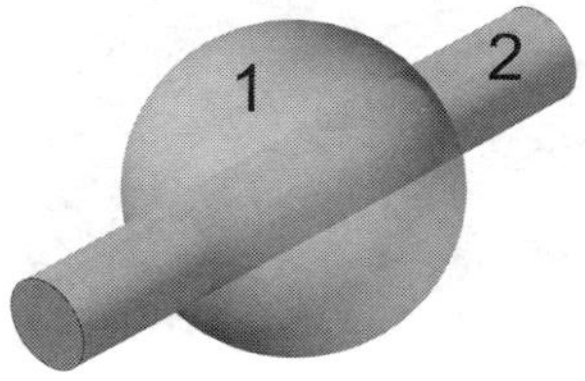

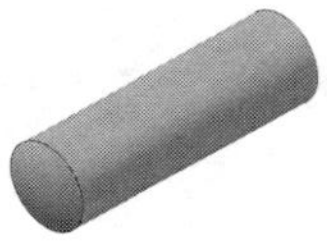

D

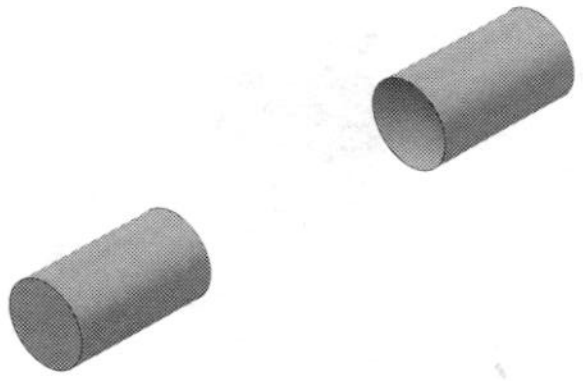

D

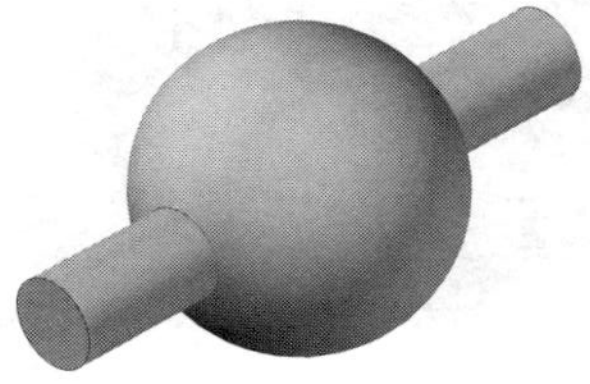

B

3.

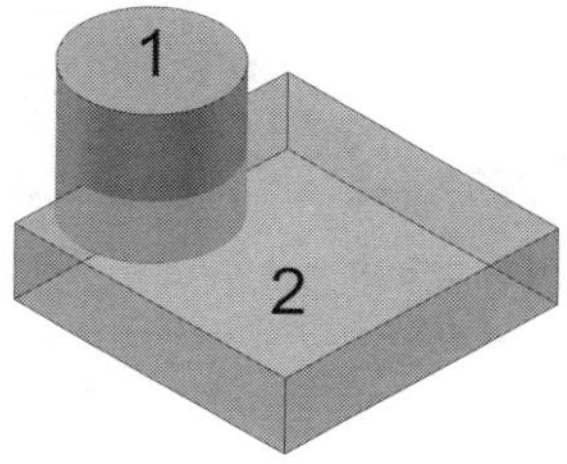

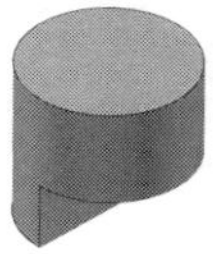

C

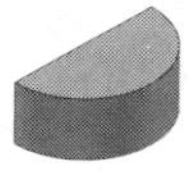

D

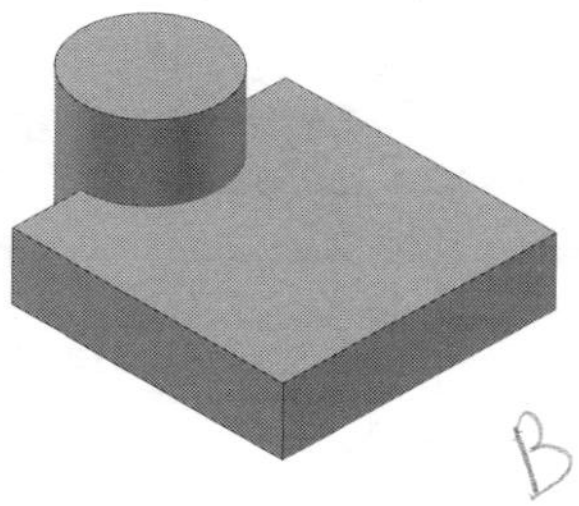

B

For the two overlapping objects shown on the left below, match the letter corresponding to the combining operation that was performed to obtain each of the three objects shown on the right.

A. Intersect   C. Object 1 cuts Object 2
B. Join   D. Object 2 cuts Object 1

1.

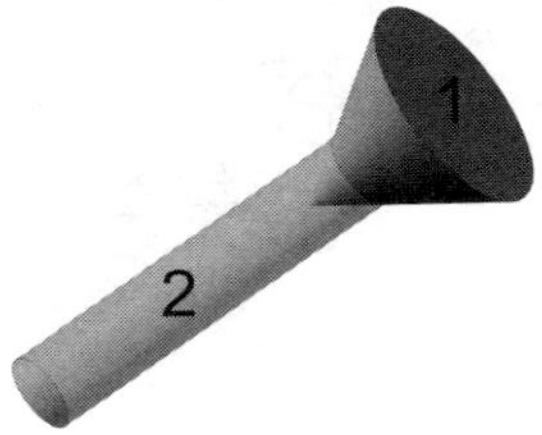

B _____

D _____

_A_ _____

2.

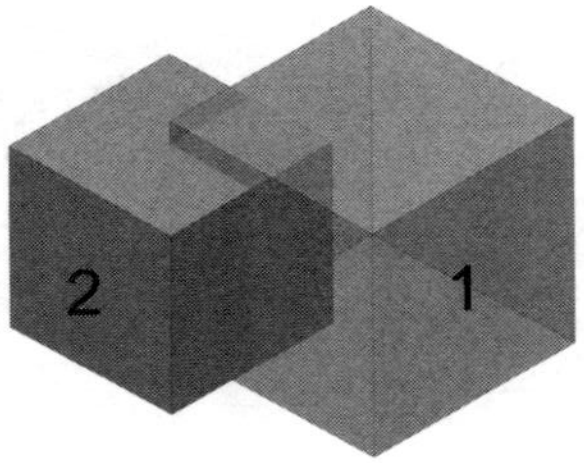

D _____

C _____

B _____

3.

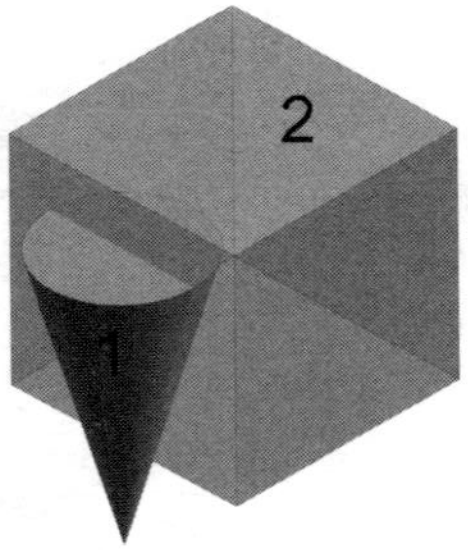

D _____

B _____

C _____

Name: Section: Date:

Class:

Developing Spatial Thinking

Grade:

Page
ob/int-06

For the two overlapping objects shown on the left below, match the letter corresponding to the combining operation that was performed to obtain each of the three objects shown on the right.

A. Intersect
B. Join
C. Object 1 cuts Object 2
D. Object 2 cuts Object 1

1.

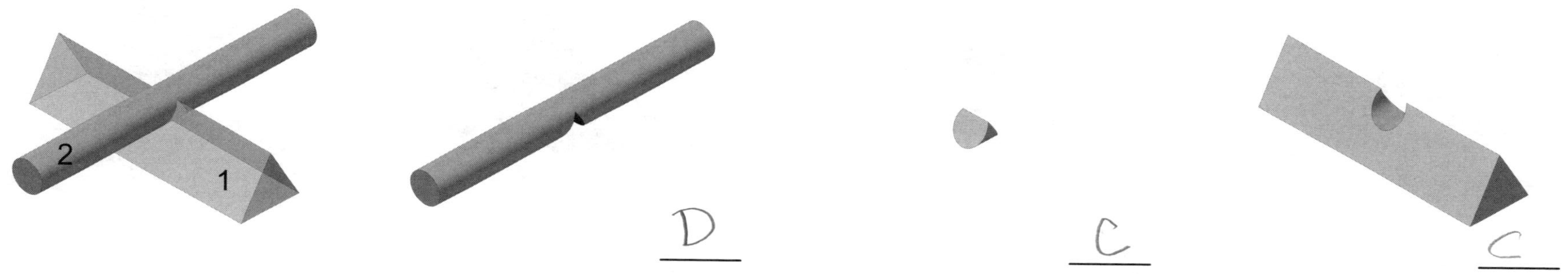

D

C

C

2.

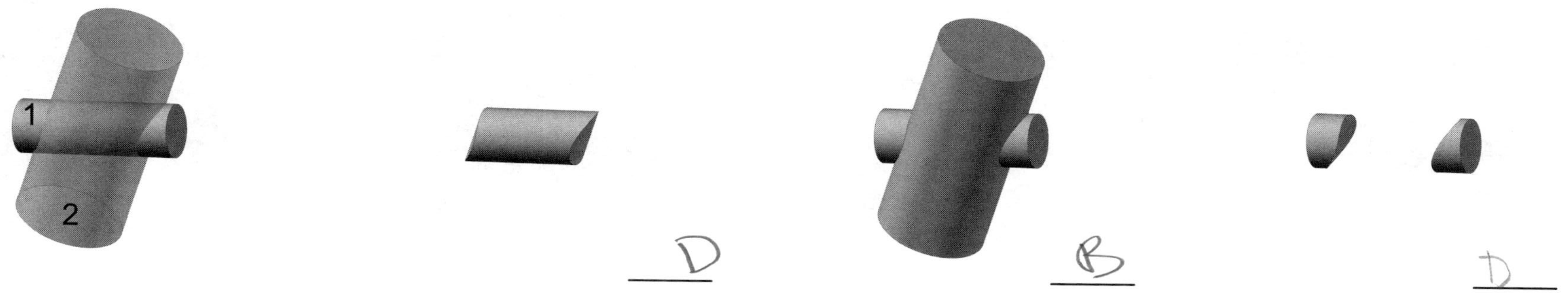

D

B

D

3.

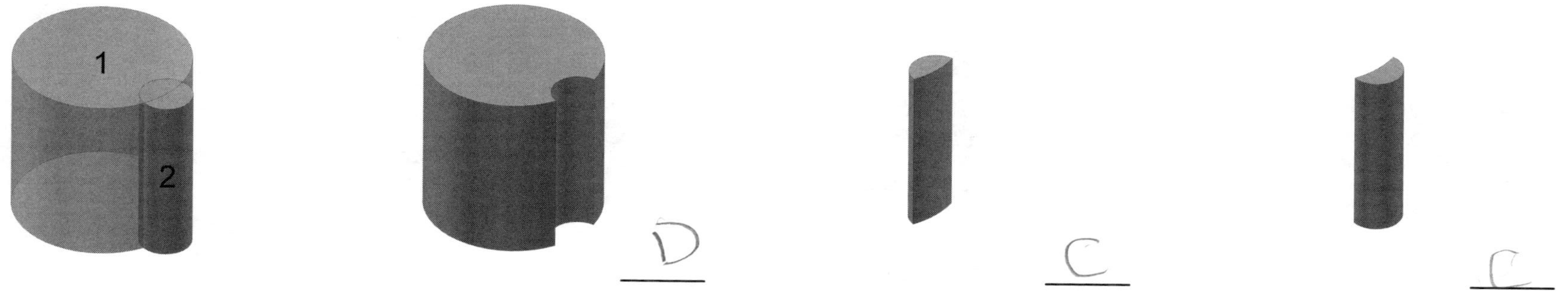

D

C

C

For the two overlapping objects shown on the left below, match the letter corresponding to the combining operation that was performed to obtain each of the three objects shown on the right.

A. Intersect
B. Join
C. Object 1 cuts Object 2
D. Object 2 cuts Object 1

1.

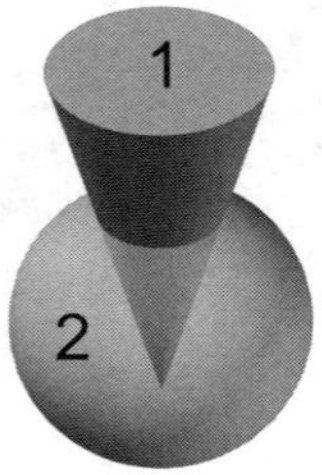

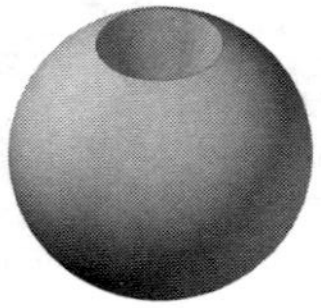

2.

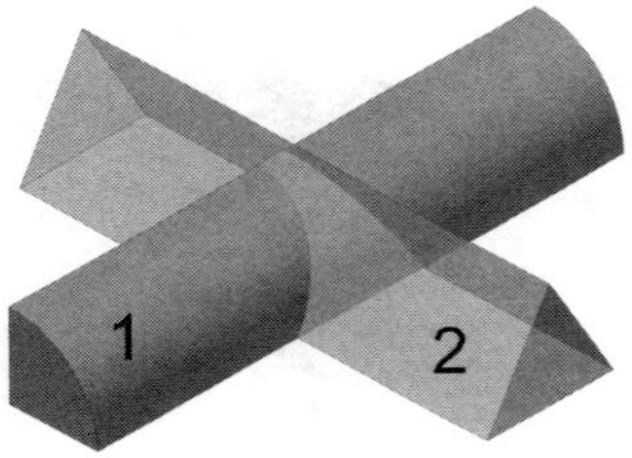

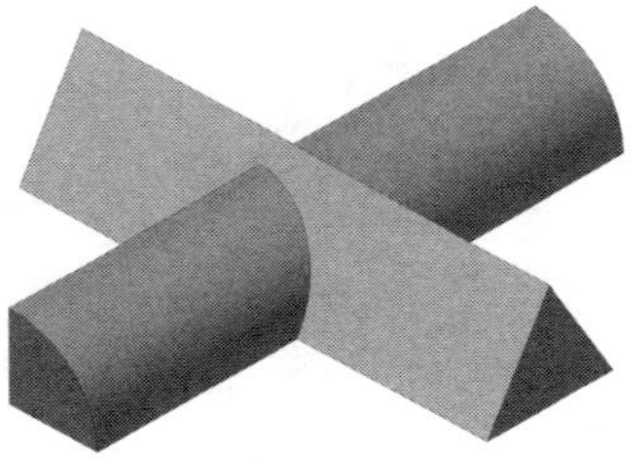

3.

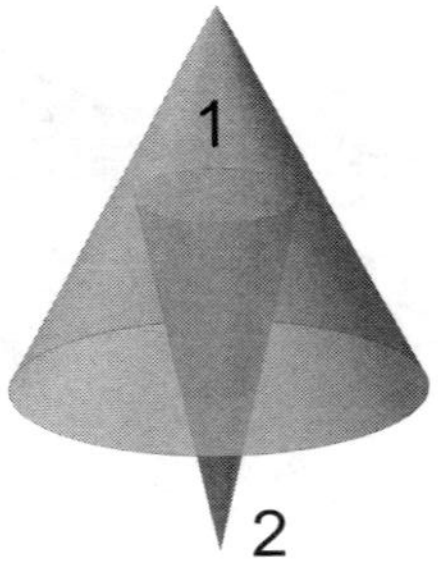

Name: Section: Date:

Class:

Developing Spatial Thinking

Grade:

Page
ob/int—08

For the two overlapping objects shown on the left below, circle the letter corresponding to the correct volume of interference.

1.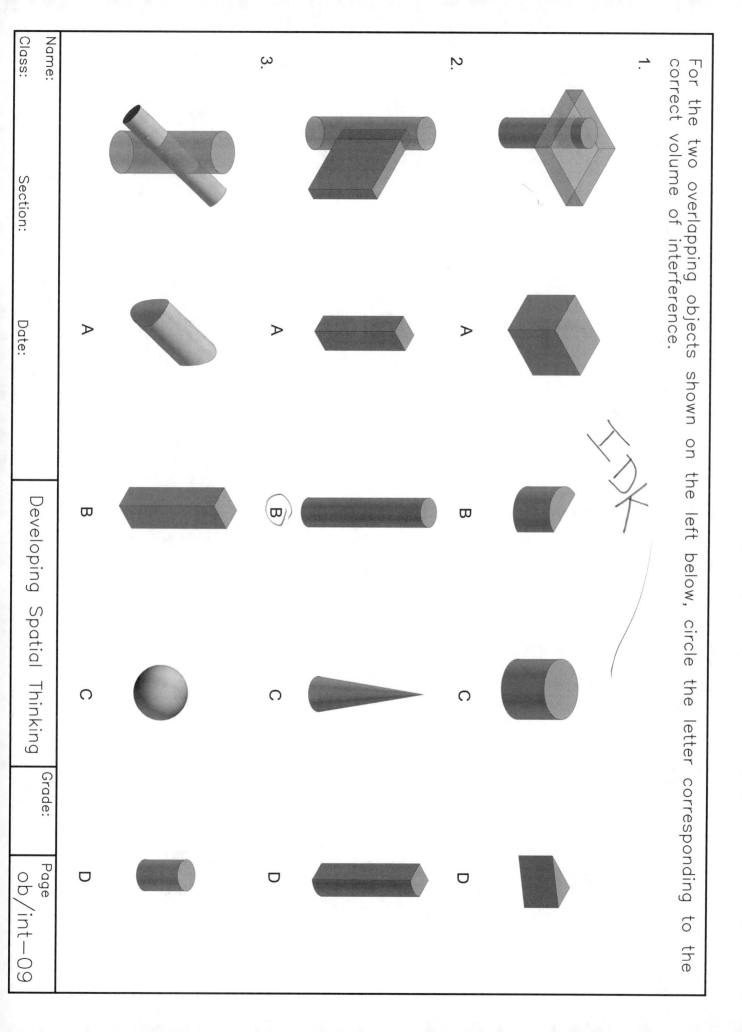

A

B

C

D

2.

A

B

C

D

3.

A

B

C

D

IDK

Name:

Class:

Section:

Date:

Developing Spatial Thinking

Grade:

Page
ob/int—09

For the two overlapping objects shown on the left below, circle the letter corresponding to the correct volume of interference.

1.

A

B

C

D

2.

A

B

C

D

3.

A

B

C

D

Name:
Class:

Section:

Date:

Developing Spatial Thinking

Grade:

Page
ob/int−10

For the two overlapping objects shown on the left below, circle the letter corresponding to the correct volume of interference.

1.

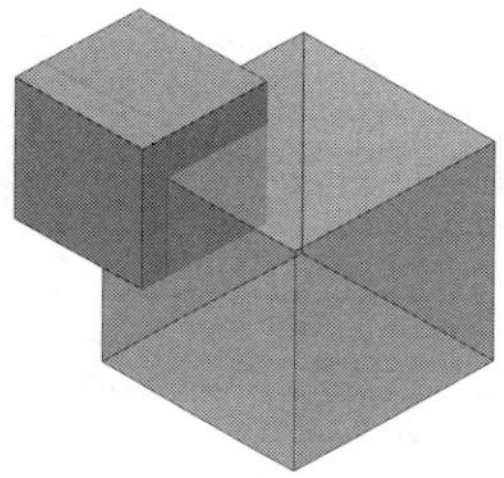

A

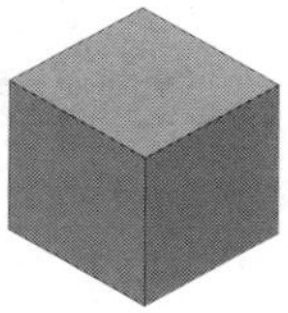

B

C

D

2.

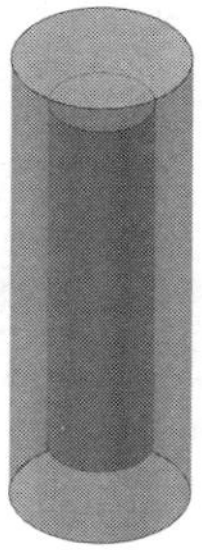

A

B

C

D

3.

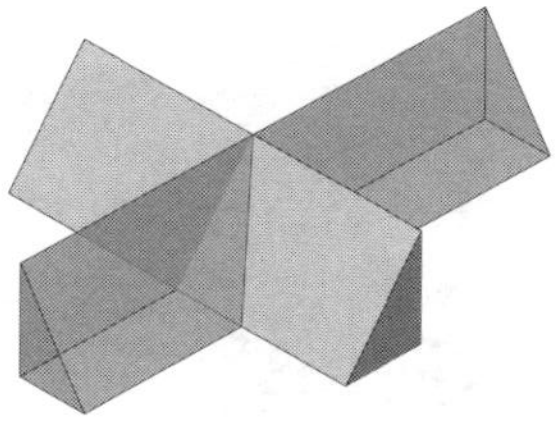

A

B

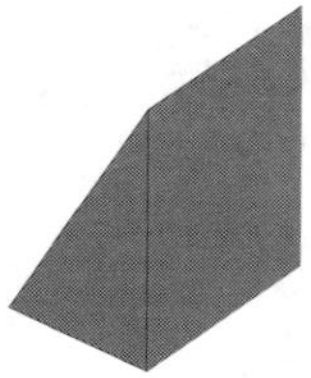

C

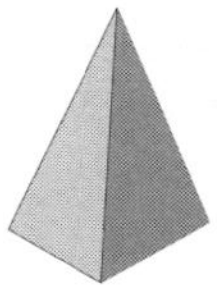

D

For the two overlapping objects shown on the left below, circle the letter corresponding to the correct volume of interference.

1.

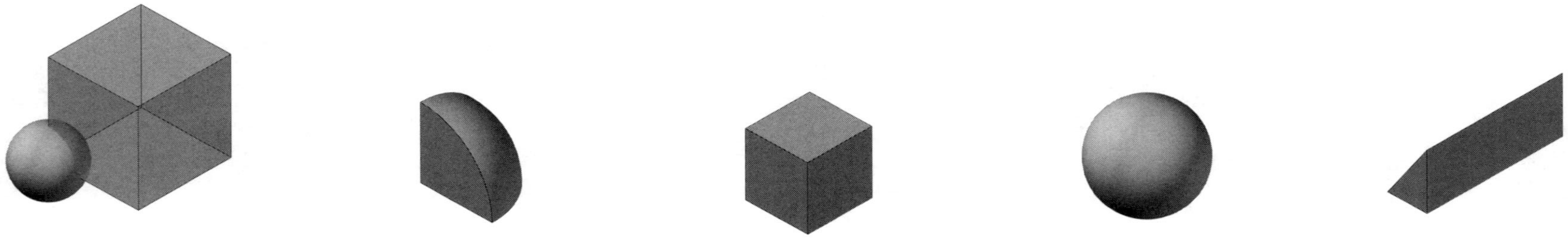

A

B

C

D

2.

A

B

C

D

3.

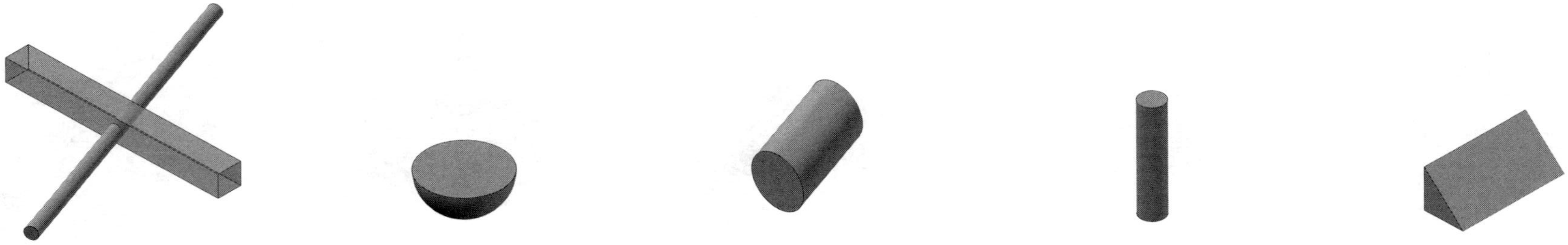

A

B

C

D

Name:

Class:

Section:

Date:

Developing Spatial Thinking

Grade:

Page
ob/int-12

Darken all edges of the resulting composite solid that obtained by performing the assigned operations.

1. A joined with B

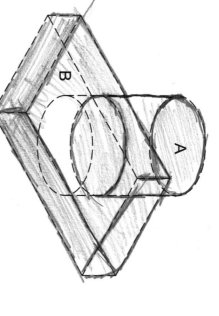

2. B cuts A

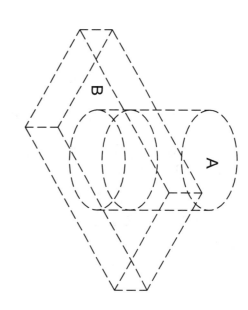

3. Intersection of A and B

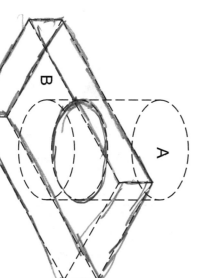

4. A cuts B

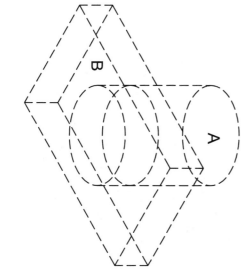

Darken all edges of the resulting composite solid that obtained by performing the assigned operations.

1. A joined with B

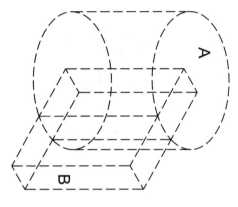

2. B cuts A

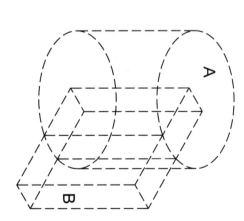

3. Intersection of A and B

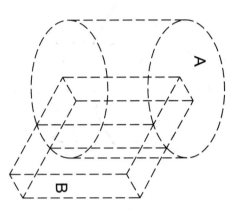

4. A cuts B

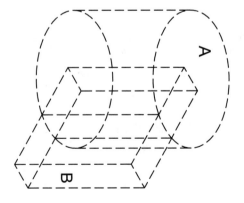

Name: Section: Date:

Class:

Developing Spatial Thinking

Grade:

Page
ob/int−14

Darken all edges of the resulting composite solid that obtained by performing the assigned operations.

1. A joined with B

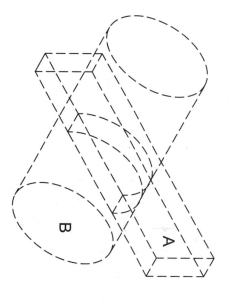

2. B cuts A

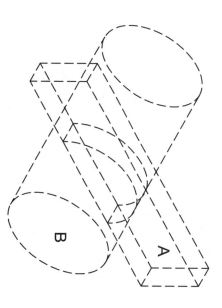

3. Intersection of A and B

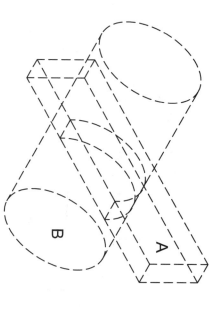

4. A cuts B

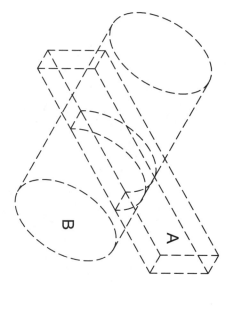

Name:

Class:

Section:

Date:

Developing Spatial Thinking

Grade:

Page
ob/int—15

Darken all edges of the resulting composite solid that obtained by performing the assigned operations.

1. A joined with B

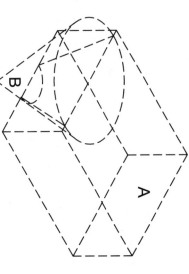

2. B cuts A

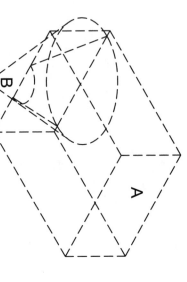

3. Intersection of A and B

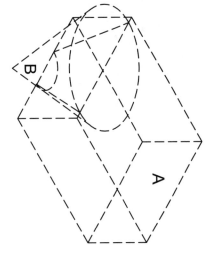

4. A cuts B

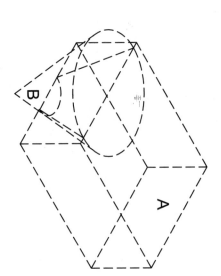

Name:

Class:

Section:

Date:

Developing Spatial Thinking

Grade:

Page
ob/int−16

Isometric Drawings & Coded Plans

Isometric views are useful for showing a 3-D object on a 2-D sheet of paper. An isometric view is defined as if you were looking down a diagonal of a cube on the object.

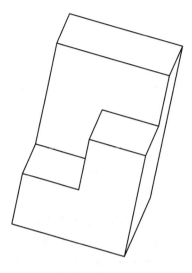

3-D Object

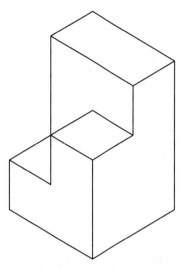

Isometric View of Object

Isometric Dot Paper is used as an aid in making isometric sketches. It consists of a grid of dots that are arranged equidistant from one another. The lines connecting the dots meet at an angle of 120 degrees with respect to one another. Isometric Grid Paper is similar to Isometric Dot Paper except that the dots are connected to form a grid on the page.

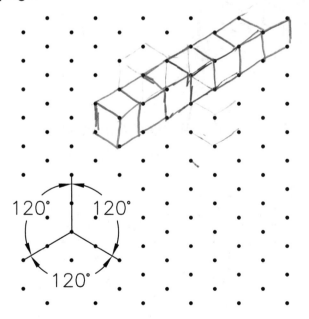

Isometric Dot Paper

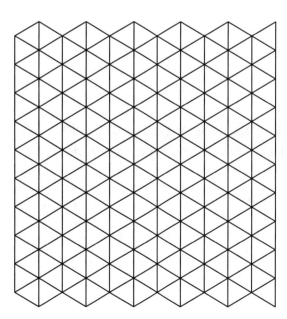

Isometric Grid Paper

When making an isometric drawing of an object created from blocks, do not show each individual cube on the object. Show only the visible surfaces and edges. An edge exists where two surfaces intersect.

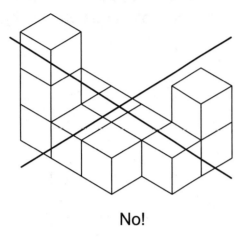

No!

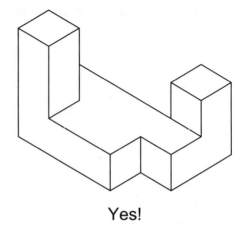

Yes!

Coded Plans are used to represent a 3-D object on a 2-D sheet of paper. Each number on the coded plan represents the height of the stack of cubes at that location. Coded plans are built up from the page to the heights specified. Corners on the coded plans are marked to indicate the viewpoint of the observer with respect to the object.

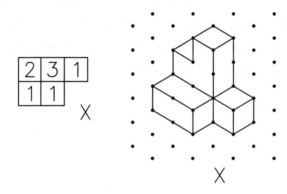

Different corner views of the object can be drawn. Notice that the object appears differently depending on your viewpoint.

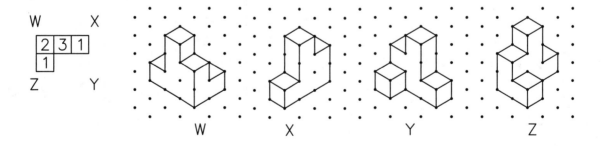

When sketching a corner view of an object, sketch one surface at a time until the view is complete.

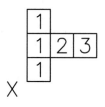

X

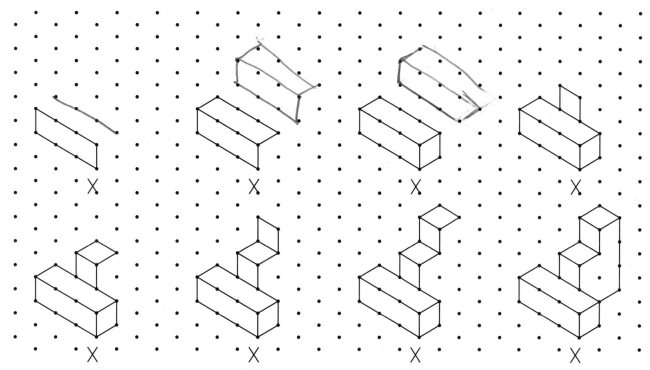

Circle the letter beneath the isometric sketch of the object that corresponds to the coded plan shown on the left.

1.

2	1
2	2
1	1

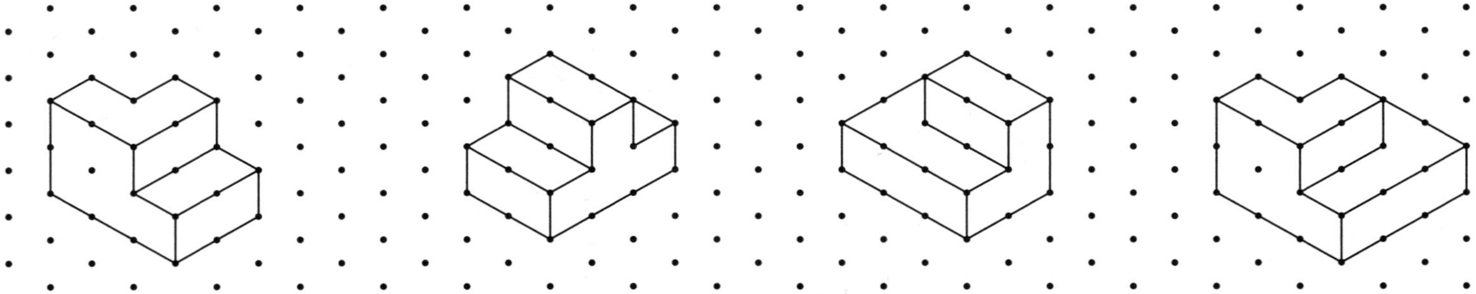

A   B   C   D

2.

3	1
2	1
1	1

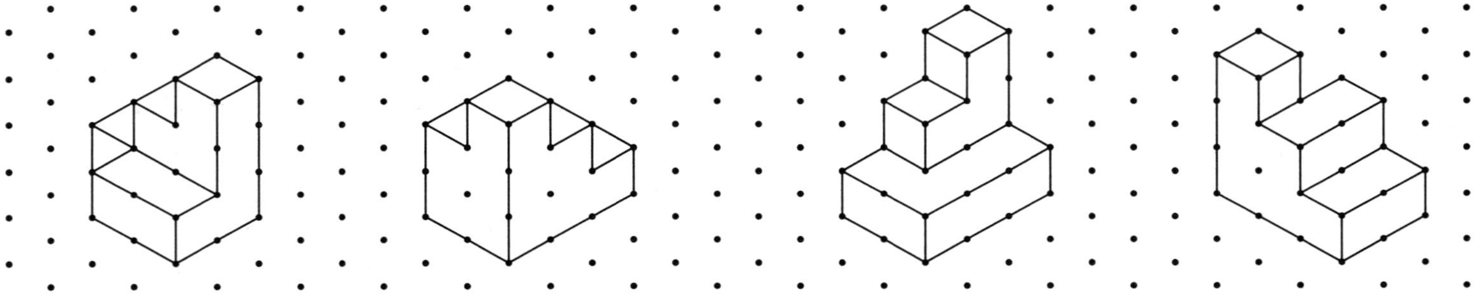

A   B   C   D

3.

3	2	
1	2	
	1	

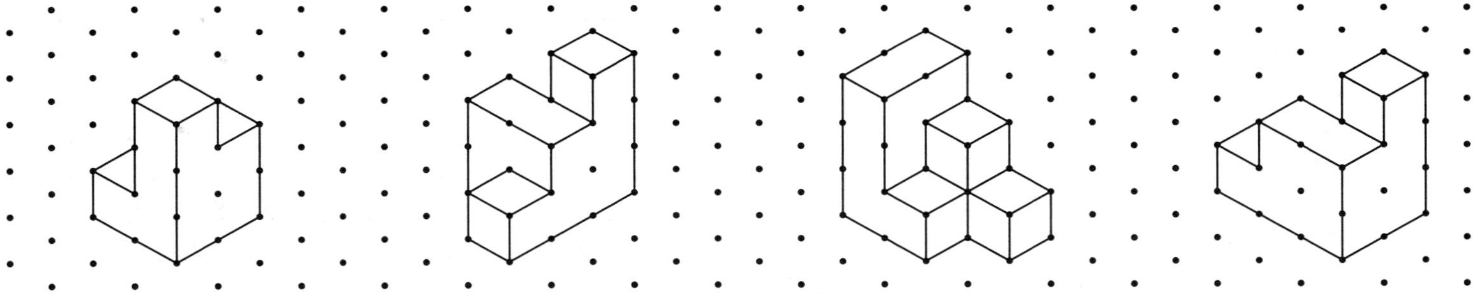

A   B   C   D

Name:

Class:

Section:

Date:

Developing Spatial Thinking

Grade:

Page
iso—01

Circle the letter beneath the isometric sketch of the object that corresponds to the coded plan shown on the left.

1.

2	
2	1
1	1

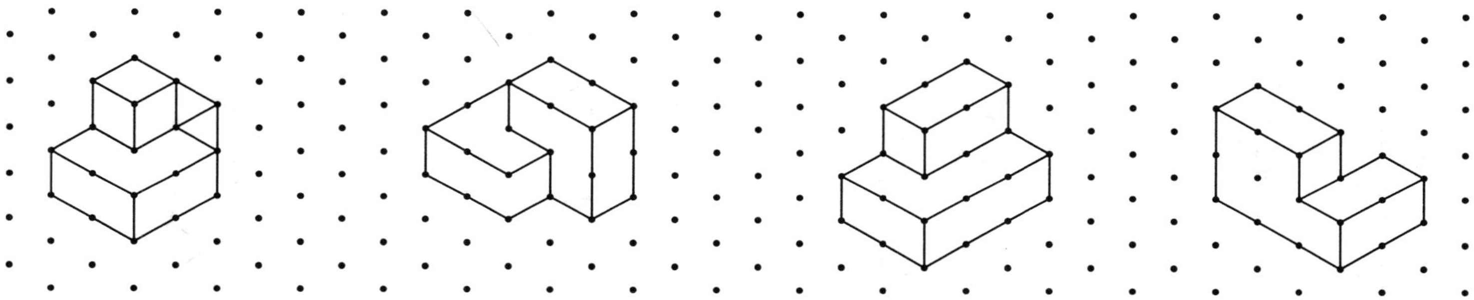

A   B   C   D

2.

1	1
1	2
3	2

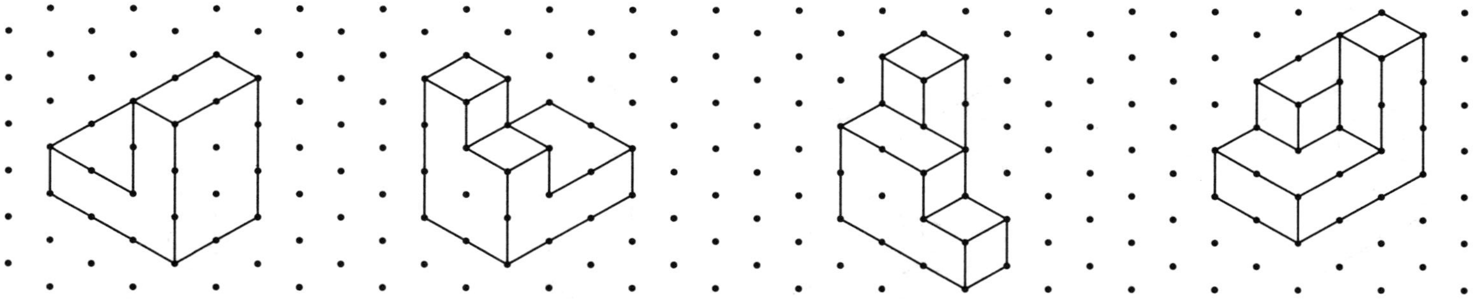

A   B   C   D

3.

	2	3
1	2	1
1		

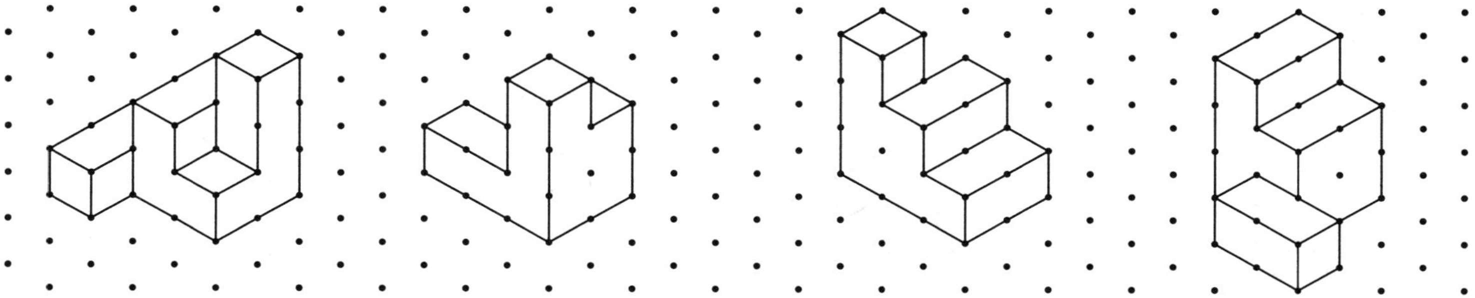

A   B   C   D

Name: Section: Date:

Class:

Developing Spatial Thinking

Grade:

Page
iso—02

Circle the letter beneath the isometric sketch of the object that corresponds to the coded plan shown on the left.

1.

3	1	2
2	1	
	1	

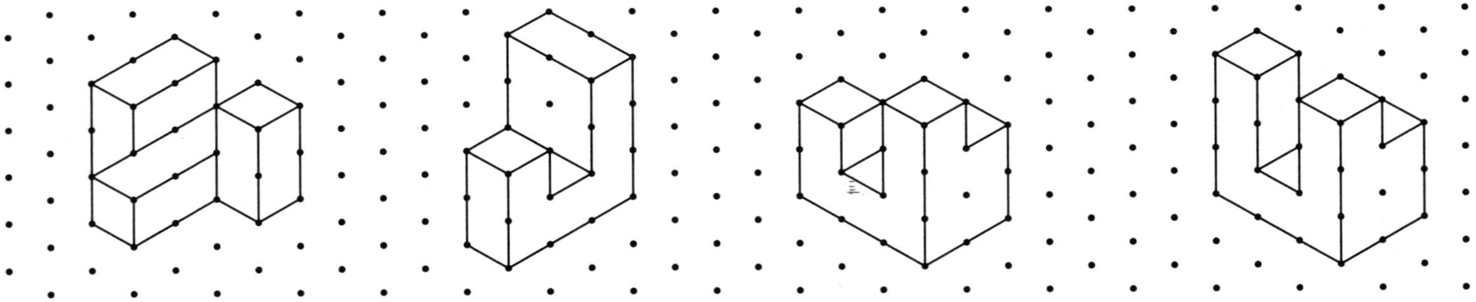

A B C D

2.

2	3	
	2	1
	1	1

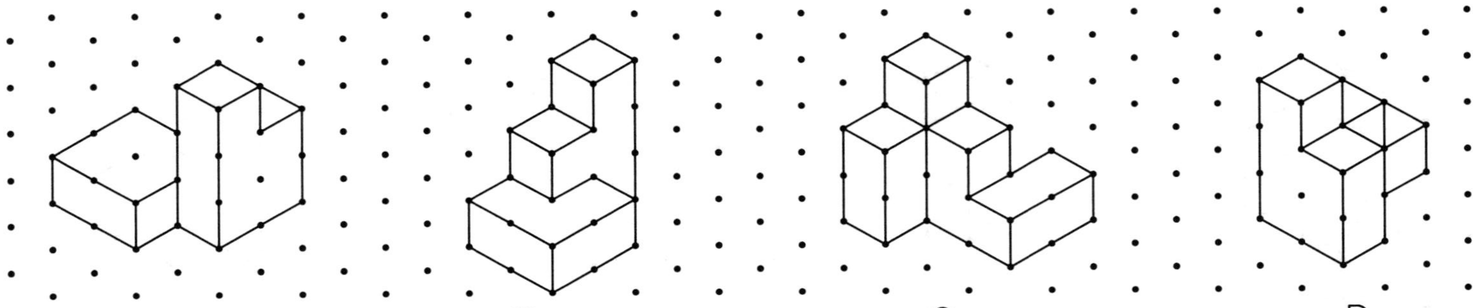

A B C D

3.

	1
3	2
1	

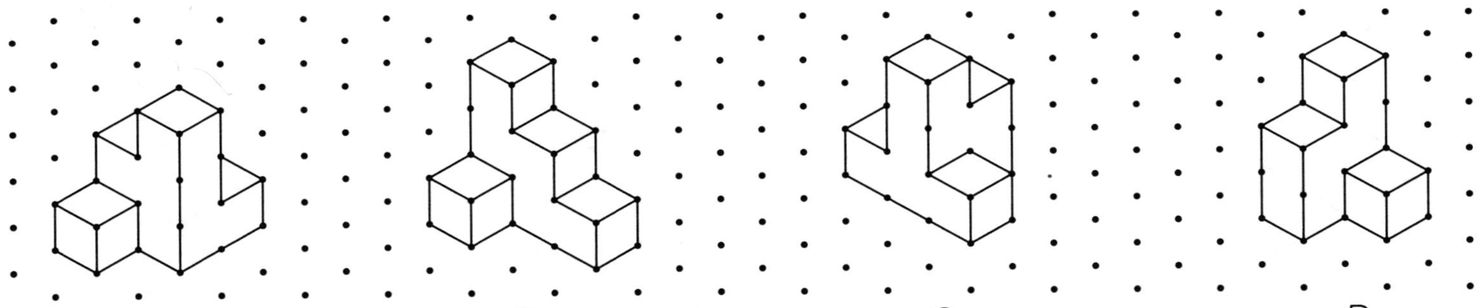

A B C D

Name:
Class: Section: Date:

Complete the coded plan for the object shown in an isometric sketch on the right.

1.

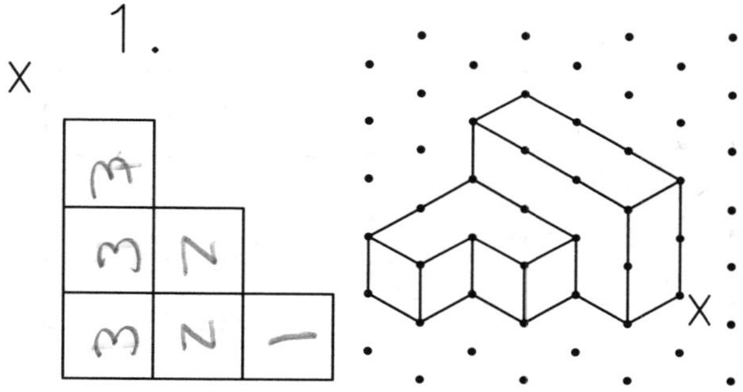

4	3	3
1	2	2

x   x

4.

x   x

2.

3	2
3	2
1	3

x

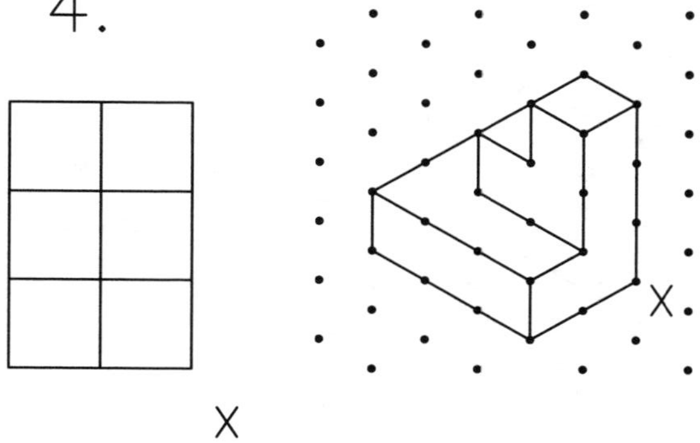

5.

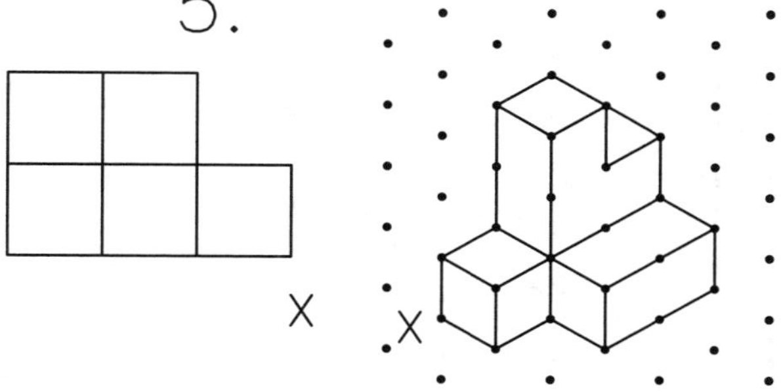

x

3.

3	2
3	2
1	3

x

6.

x

Name:
Class:

Section:

Date:

Developing Spatial Thinking

Grade:

Page
iSO—04

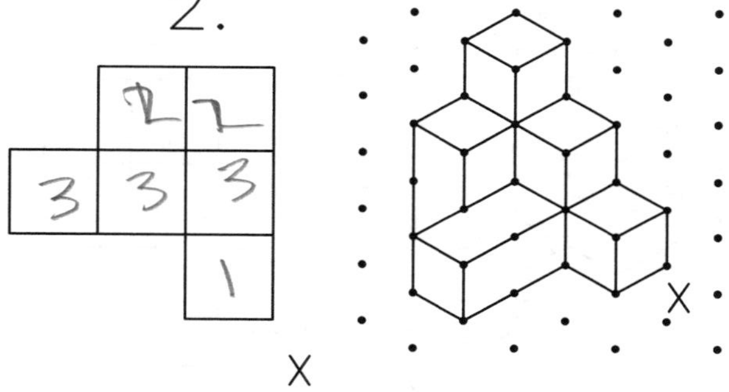

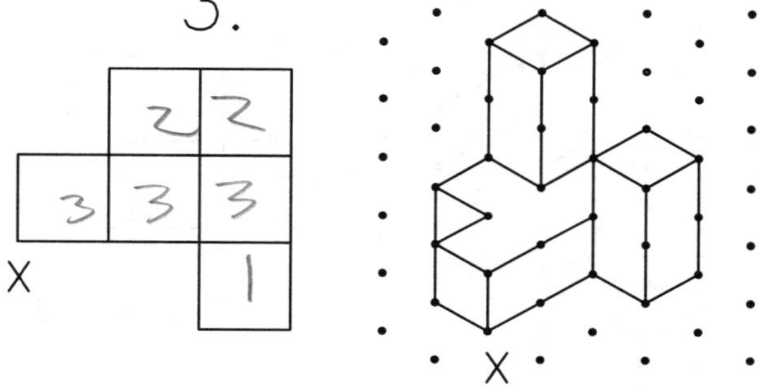

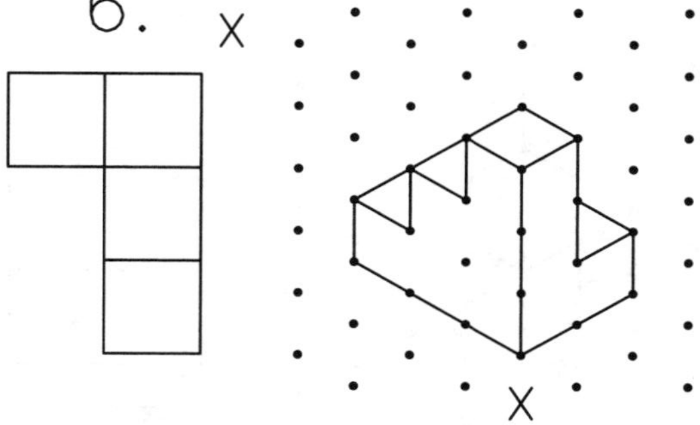

Complete the coded plan for the object shown in an isometric sketch on the right.

1.   ×

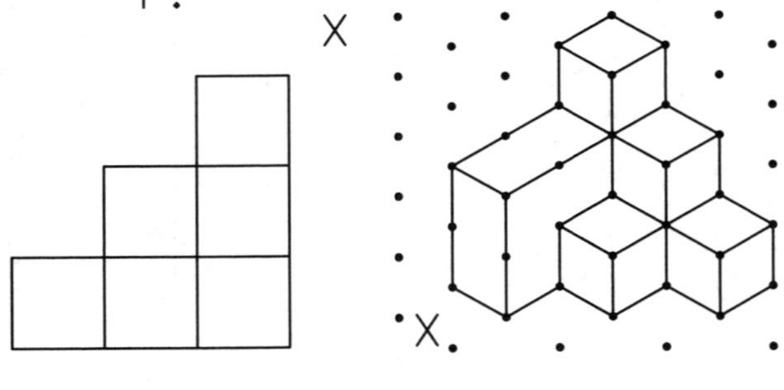

×

×

2.

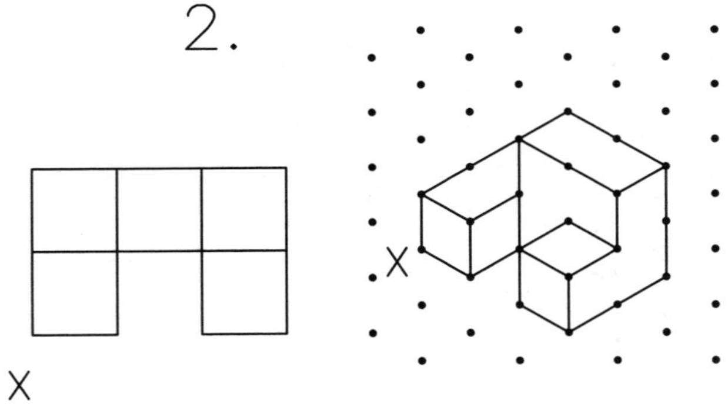

×

×

3.

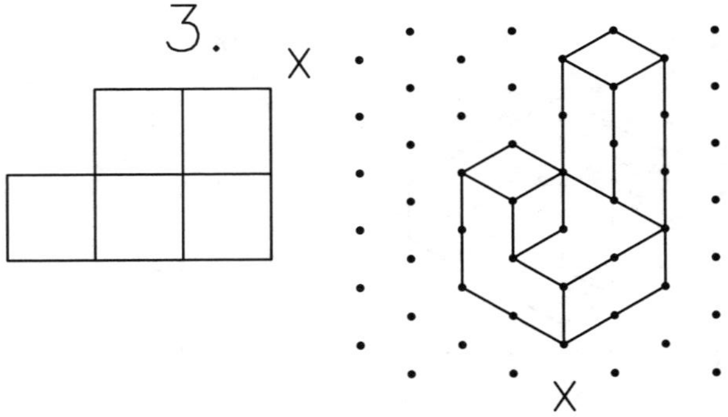

×

×

4.   ×

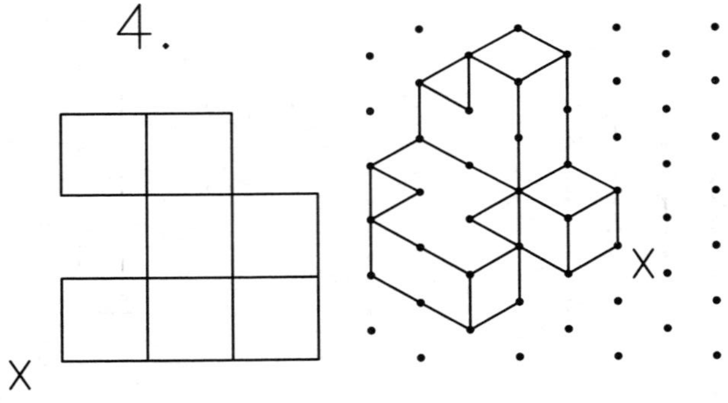

×

×

5.   ×

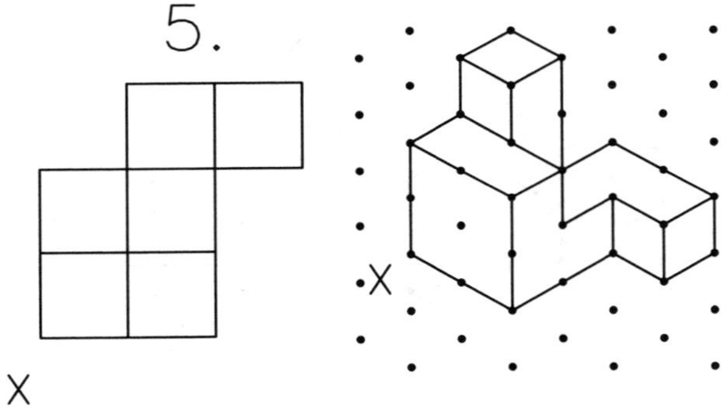

×

6.   ×

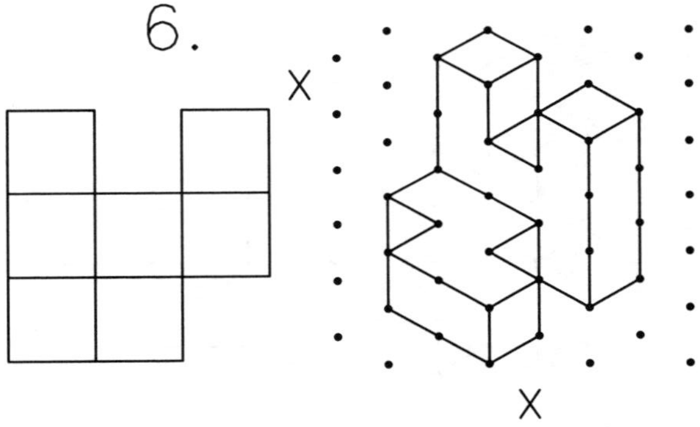

×

Complete the coded plan for the object shown in an isometric sketch on the right.

1.

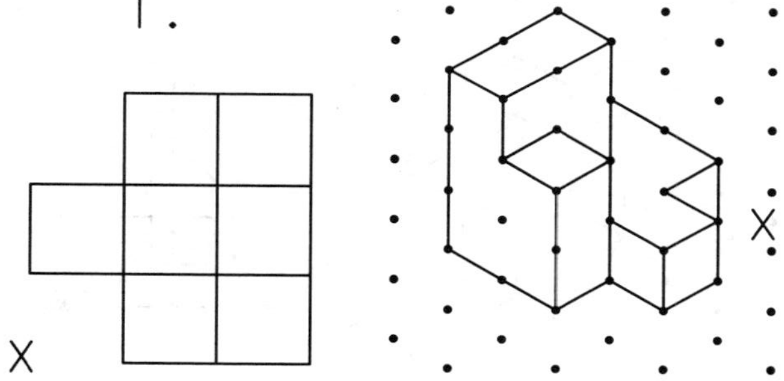

2.

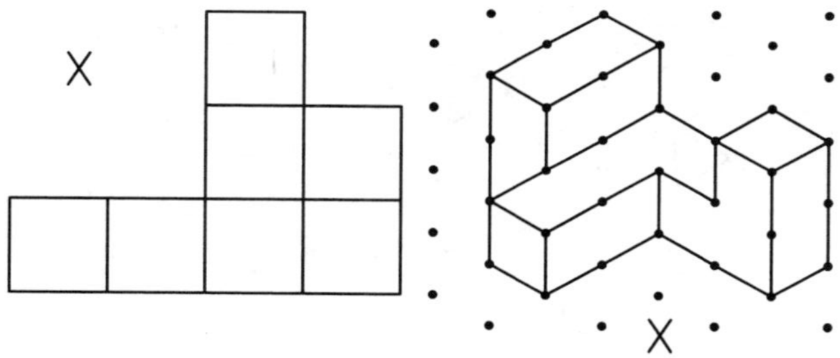

3.

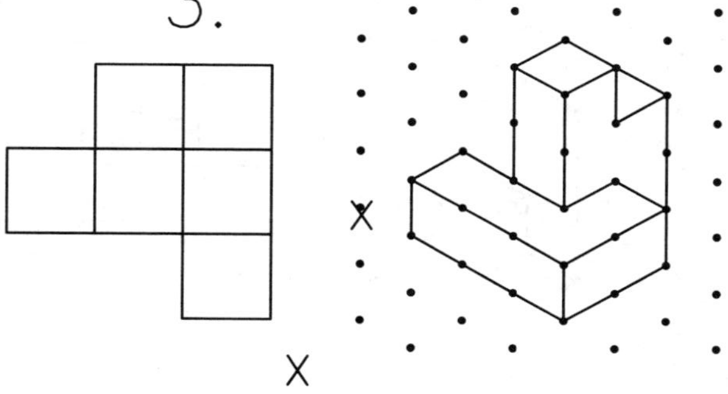

4.

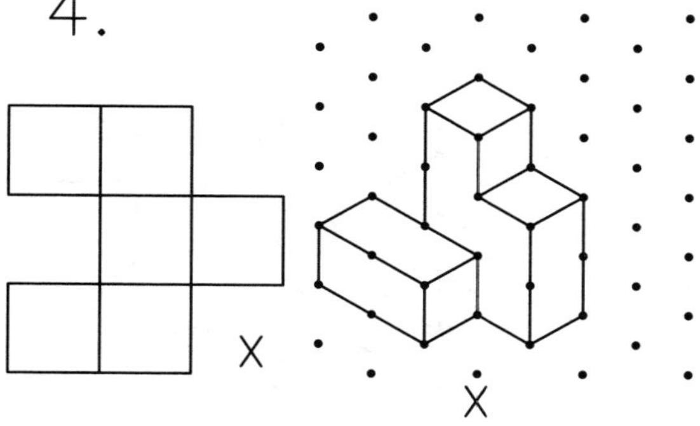

5.

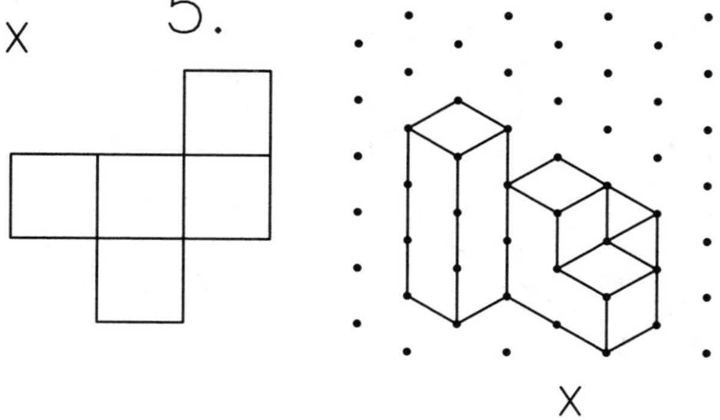

6.

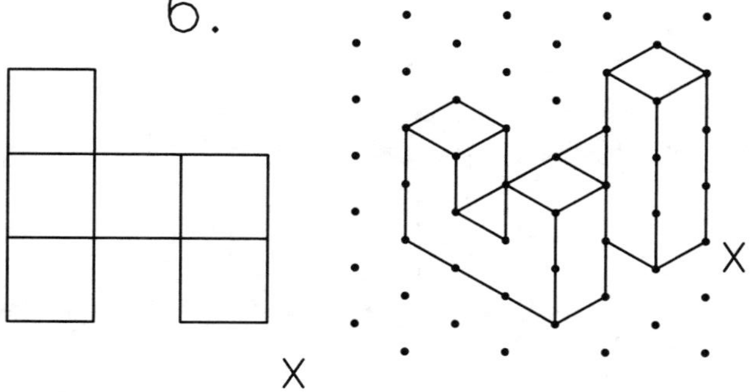

Name:
Class:
Section:
Date:

Developing Spatial Thinking

Grade:

Page
iso—06

Circle the letter on the coded plan (W, X, Y, or Z) that corresponds to the given isometric sketch.

1.

W   X

1	
1	1
2	3

Z   Y

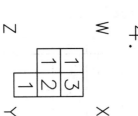

2.

W   X

	3	
1	2	2

Z   Y

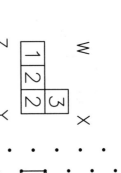

3.

W   X

		1
1	3	
2		

Z   Y

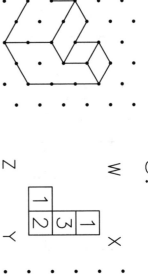

4.

W   X

	1	
1	3	
2	2	
	1	

Z   Y

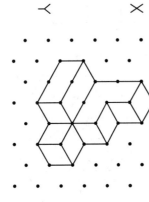

5.

W   X

1	
2	1
1	
1	

Z   Y

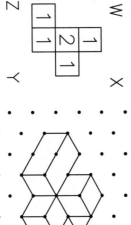

6.

W   X

2	
3	2	1

Z   Y

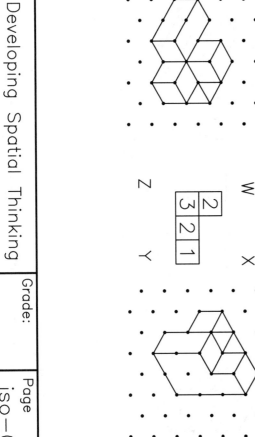

Name: Section: Date:

Class:

Developing Spatial Thinking

Grade:

Page
iSO—07

Circle the letter on the coded plan (W, X, Y, or Z) that corresponds to the given isometric sketch.

1.

W   X

1	1
3	3
2	2

Z   Y

2.

W   X

1	1
3	2
1	2

Z   Y

3.

W   X

2	1
1	1
2	

Z   Y

4.

W   X

1	1
1	2
3	
2	

Z   Y

5.

W   X

2	1
1	1
2	

Z   Y

6.

W   X

2	1
1	1
2	

Z   Y

Developing Spatial Thinking

Name:
Class:
Section:
Date:
Grade:
Page
iso—08

Circle the letter on the coded plan (W, X, Y, or Z) that corresponds to the given isometric sketch.

1.

W
1	1
1	3

1
2

Z   X
                                     Y

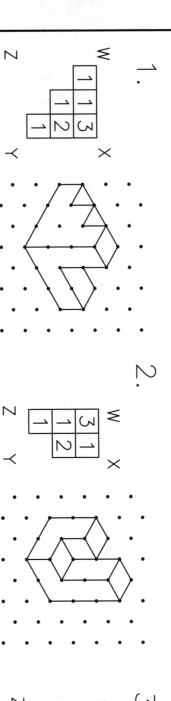

2.

W
3	1
1	2

1

Z   X
                                     Y

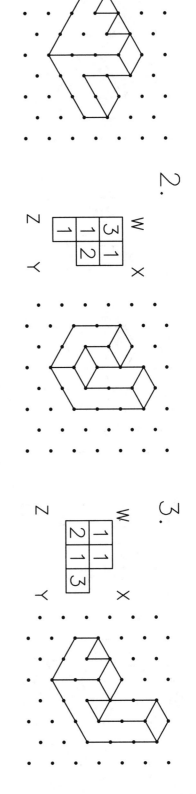

3.

W
1	1
2	1
1	3

Z   X
                                     Y

4.

W
1	2
1	3

1	1
	1

Z   X
                                     Y

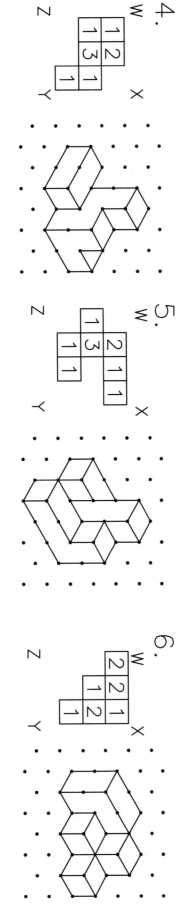

5.

W
2	1
1	3
1	1

Z   X
                                     Y

6.

W
2	2	1
	1	2
		1

Z   X
                                     Y

Name: Section: Date:

Class:

Developing Spatial Thinking

Grade:

Page
iso—09

Sketch the indicated corner view in the space provided.

1.

3		
2	1	
1	1	1

x

2.

	2
3	
1	

x

3.

2	3
1	1
1	

x

4.

1	1	3
1	2	

x

5.

2	1	
1	1	
3	2	1

x

6.

2	3	
1	1	
1	2	

x

Name:
Class:

Section:

Date:

Developing Spatial Thinking

Grade:

Page
iso—10

Sketch the indicated corner view in the space provided.

1.

3	1
1	2
1	1

x

2.

x

3	1
1	1
2	1
	1

3.

2	1
3	
2	1
1	1

x

4.

x

2	
3	
3	1	1

5.

	3
2	1
2	

x

6.

x

3	2	1
2		

Name: Section: Date:

Class:

Developing Spatial Thinking   Grade: Page
                                                     iso—11

Sketch the indicated corner view in the space provided.

1.

2	1
1	1
1	3

x

2.

1	1	
1	2	3
		1

x

3.

2	3
1	2
	1

x

4.

1	1
1	2
3	3

x

5.

1	3	2	
	1	1	1

x

6.

	3	1	2
	1	1	
		1	

x

Orthographic Drawings

Isometric views show objects from their corners. Orthographic views show the faces of the object straight on or parallel to the viewing plane.

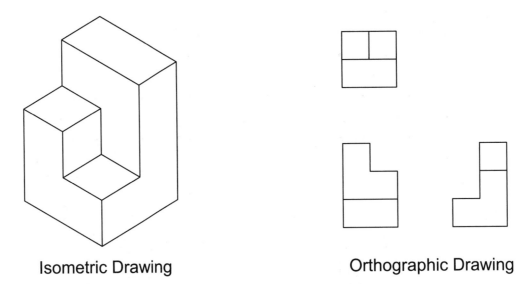

Isometric Drawing

Orthographic Drawing

In creating orthographic views, you imagine that the object is surrounded by a transparent glass cube. The edges and surfaces of the object are projected onto the panes of the glass cube and the cube is unfolded so the panes of glass lie in one plane.

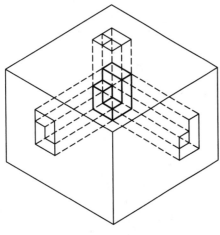

Object Surrounded by
Glass Cube

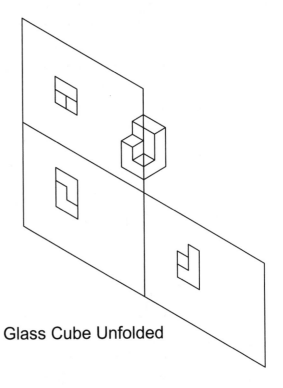

Glass Cube Unfolded

Normal surfaces are defined as being parallel to either the top, front, or side views. A normal surface is seen as a surface in the view to which it is parallel and is seen as an edge in the other views.

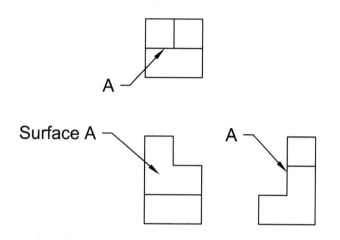

Rules in orthographic projection:

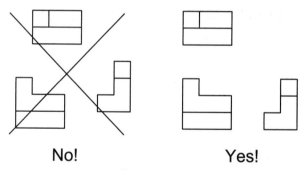

No!   Yes!

The orthographic views of an object should be aligned with one another.

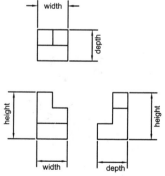

The top view shows the width and depth of an object, the front view shows the height and width, and the side view shows the height and depth.

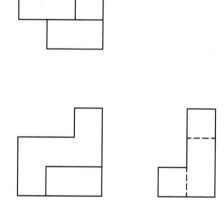

Edges of an object that are hidden from a given viewpoint are shown as dashed lines in that view.

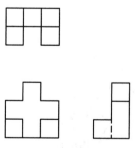

When a hidden line coincides with a solid line on a view, show only the solid line.

Sometimes only two views are required to completely describe an object. However, most times, three views are shown (top, front, and right side).

Only 2 Views Required   3 Views Required

The box method is sometimes used to create an isometric drawing from three orthographic views. Step1: Construct an isometric drawing of a box that is the same overall size as the object seen in the orthographic views. Step2: Draw the top, front, and side views on this box. Step3: Add and erase lines from the isometric drawing to complete it.

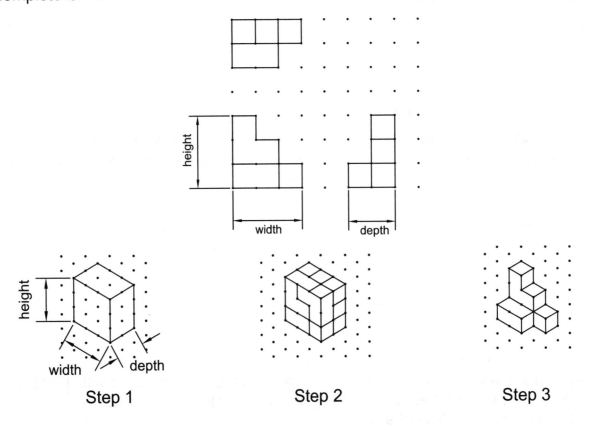

Step 1   Step 2   Step 3

An isometric view of an object is shown below along with its top and front views.
Circle the letter corresponding to the correct side view from the choices given

1.

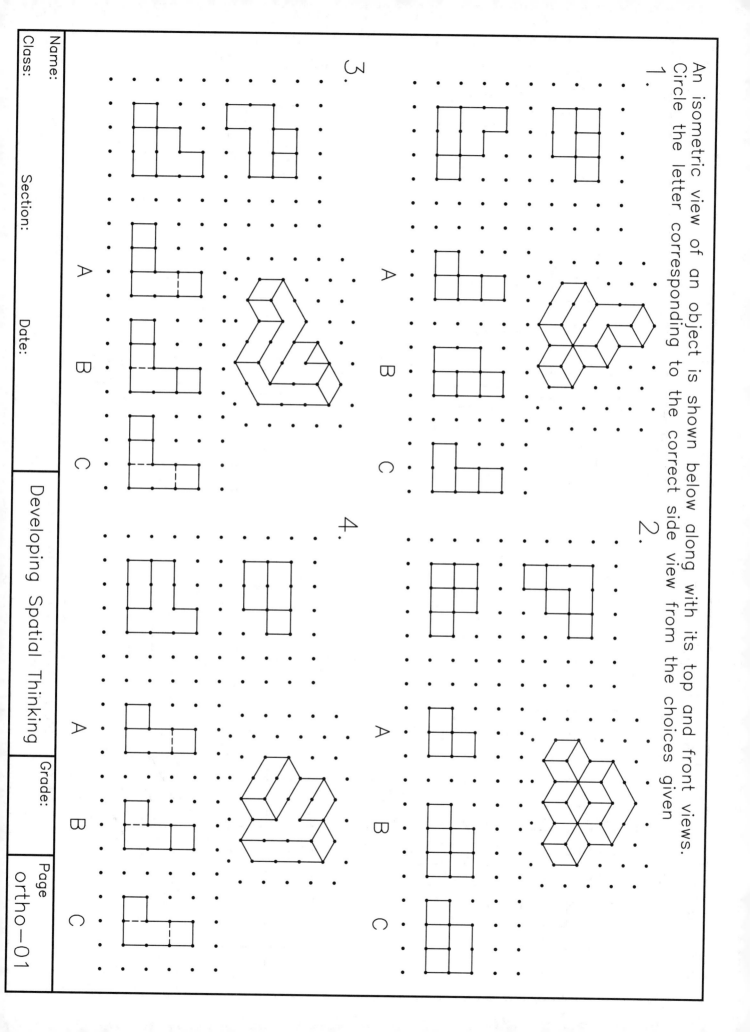

A   B   C

2.

A   B   C

3.

A   B   C

4.

A   B   C

An isometric view of an object is shown below along with its top and front views.
Circle the letter corresponding to the correct side view from the choices given

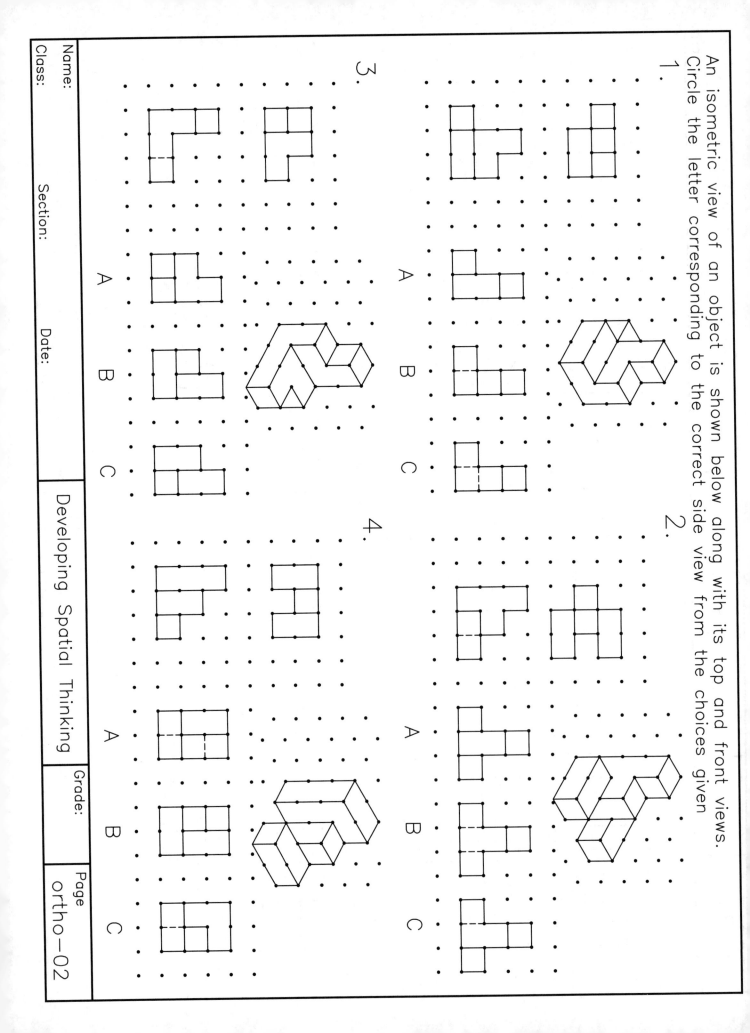

1.

A   B   C

2.

A   B   C

3.

A   B   C

4.

A   B   C

Developing Spatial Thinking   Grade: Page
                                                      ortho−02

An isometric view of an object is shown below along with its top and front views.
Circle the letter corresponding to the correct side view from the choices given

1.

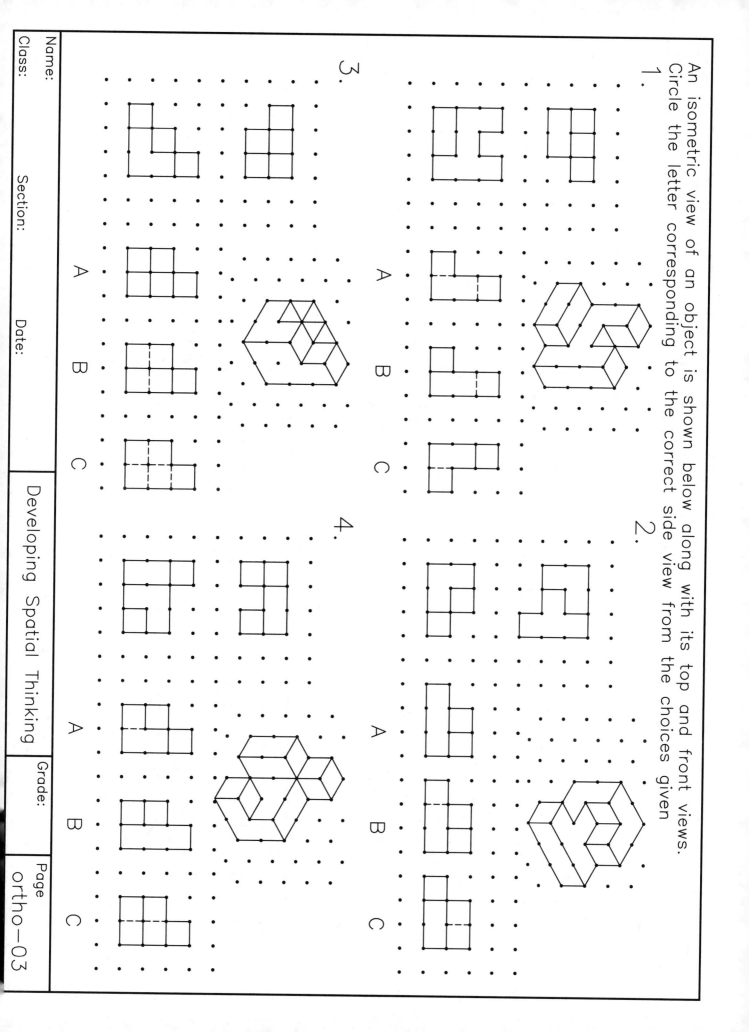

A   B   C

2.

A   B   C

3.

A   B   C

4.

A   B   C

An isometric view of an object is shown below along with it's top and front views.
Circle the letter corresponding to the correct side view from the choices given

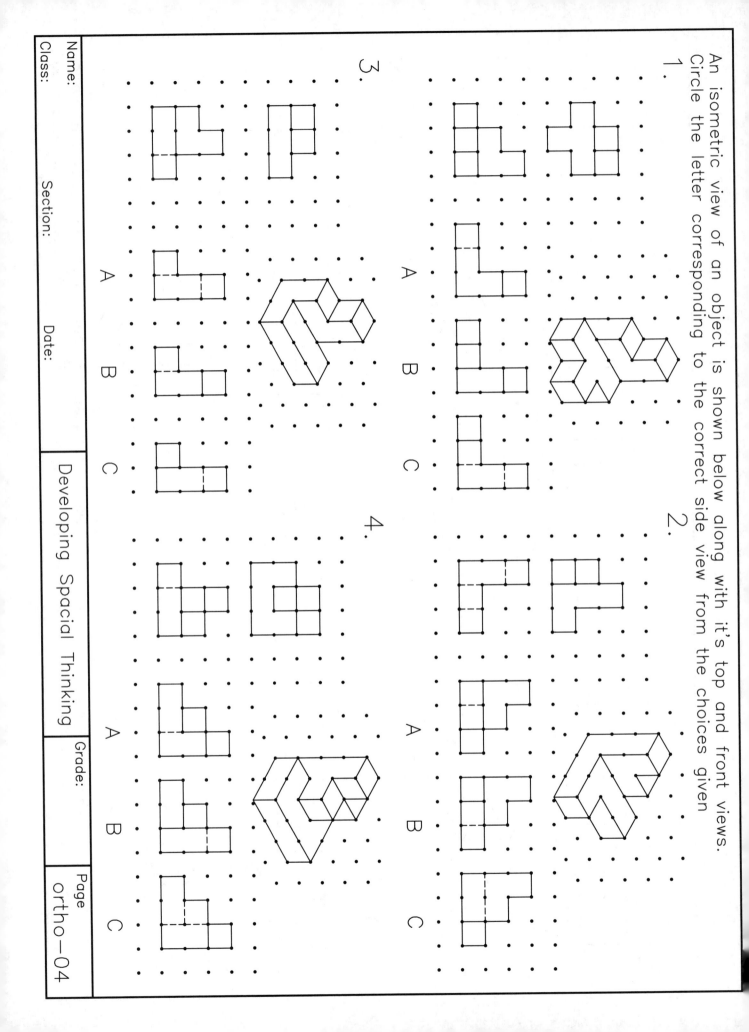

1.

A   B   C

2.

A   B   C

3.

A   B   C

4.

A   B   C

For the object shown in orthographic projection on the left,
circle the letter of the correct corresponding isometric view.

1.

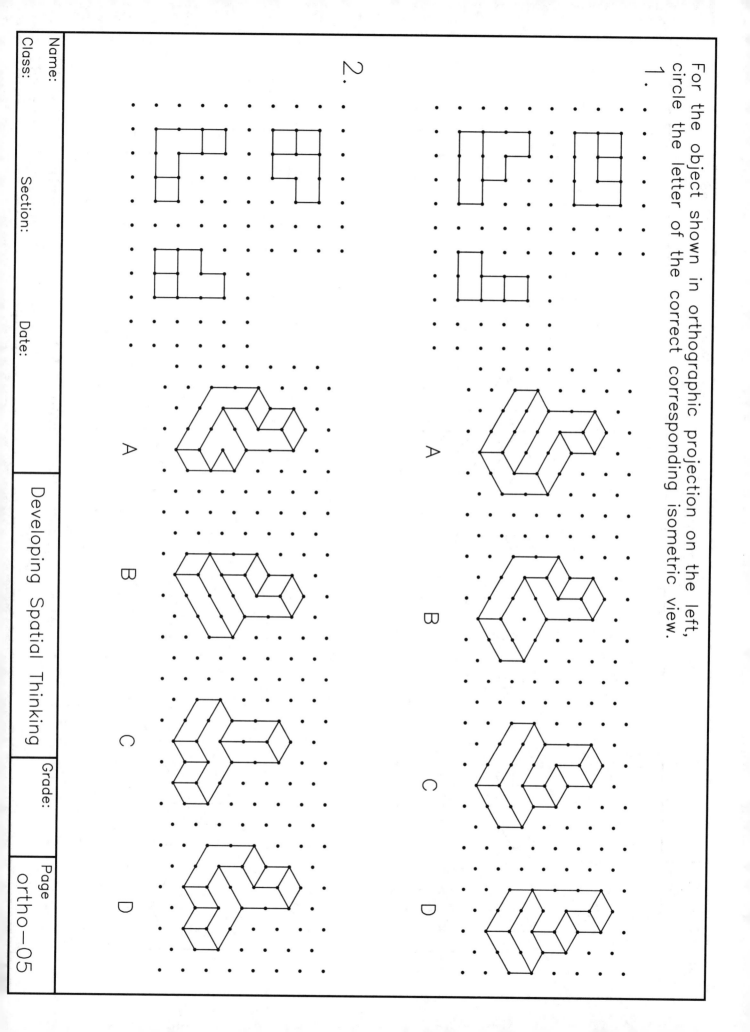

A   B   C   D

2.

A   B   C   D

Name:

Class:

Section:

Date:

Developing Spatial Thinking

Grade:

Page
ortho-05

For the object shown in orthographic projection on the left, circle the letter of the correct corresponding isometric view.

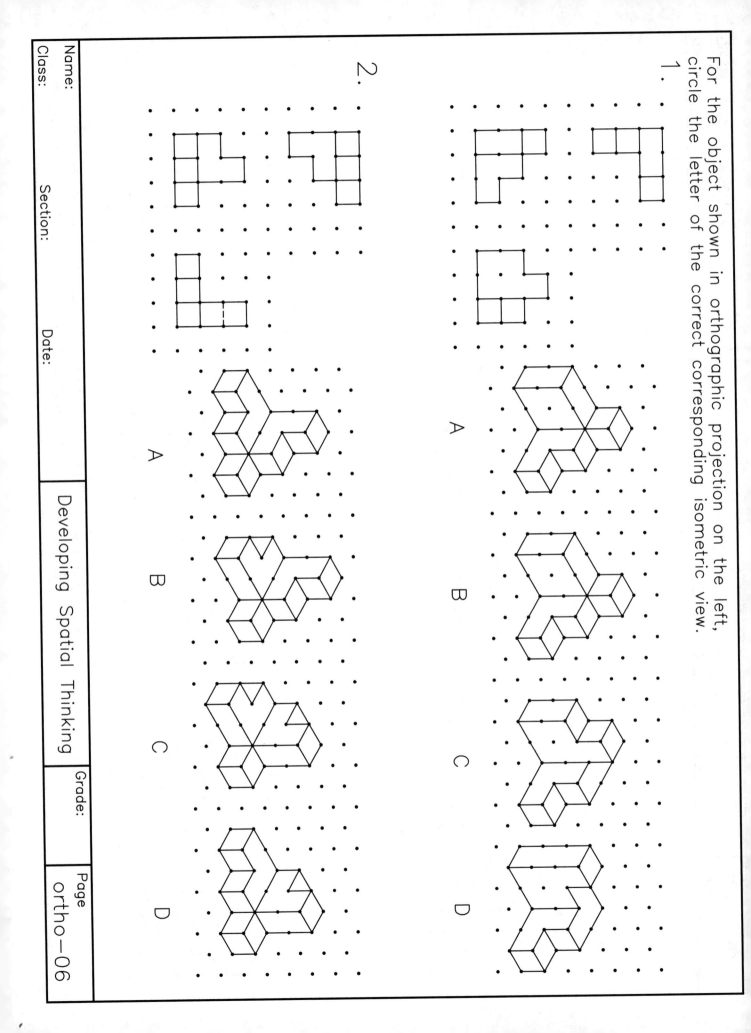

1.

A   B   C   D

2.

A   B   C   D

For the object shown in orthographic projection on the left,
circle the letter of the correct corresponding isometric view.

1.

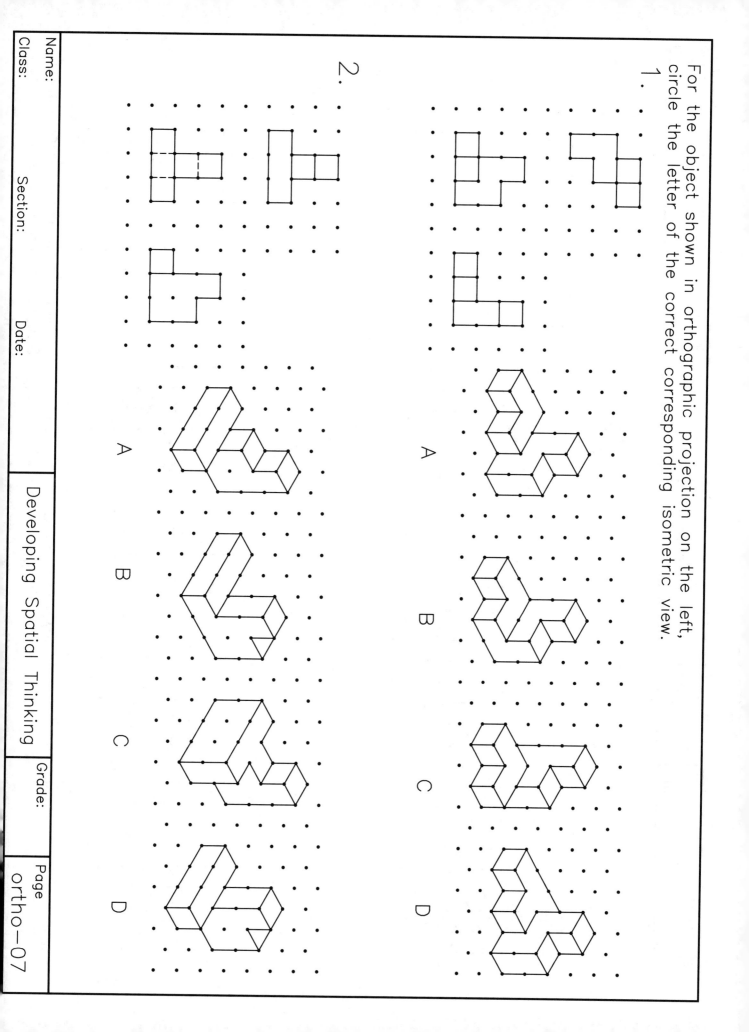

A   B   C   D

2.

A   B   C   D

For the object shown in orthographic projection on the left, circle the letter of the correct corresponding isometric view.

1.

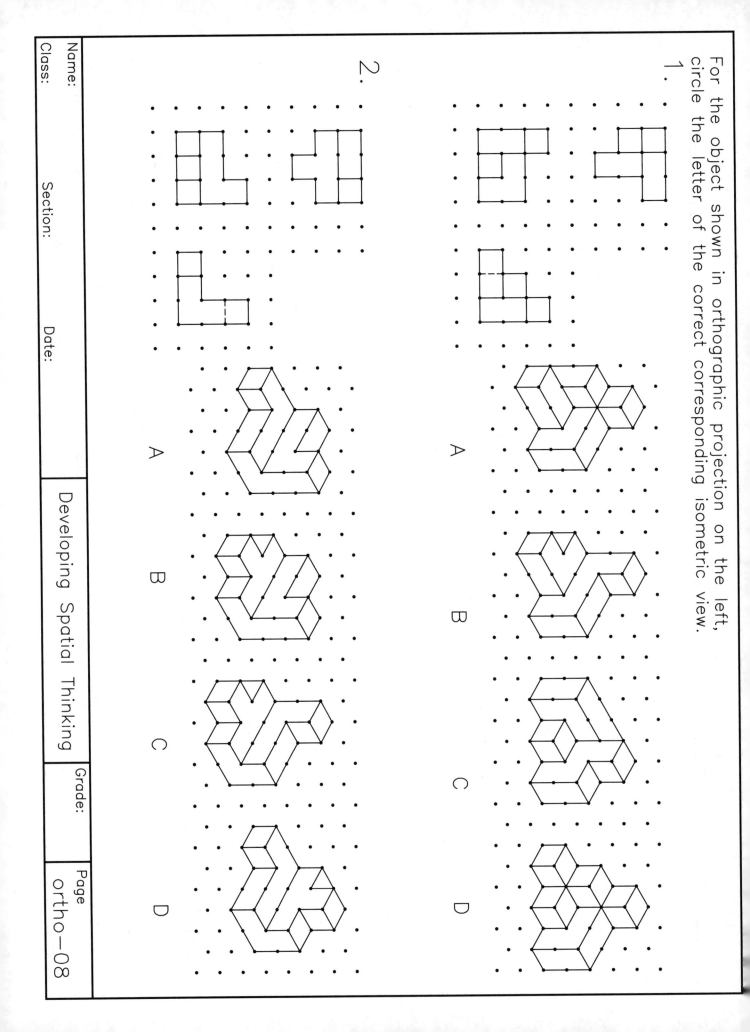

A   B   C   D

2.

A   B   C   D

Name: _____ Section: _____ Date: _____

Class: _____

Developing Spatial Thinking

Grade: _____

Page
ortho—08

For the object shown in orthographic projection on the left,
circle the letter of the correct corresponding isometric view.

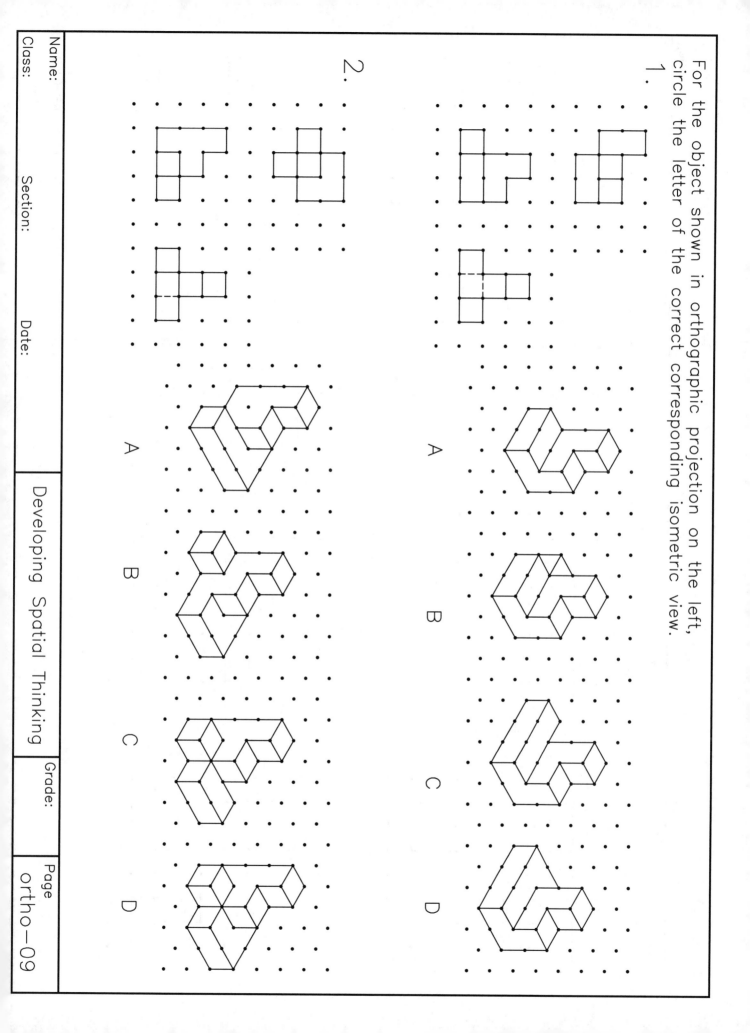

1.

A   B   C   D

2.

A   B   C   D

Name:

Class:

Section:

Date:

Developing Spatial Thinking

Grade:

Page
ortho-09

Circle the letter corresponding to the correctly aligned set of orthographic views.

1.

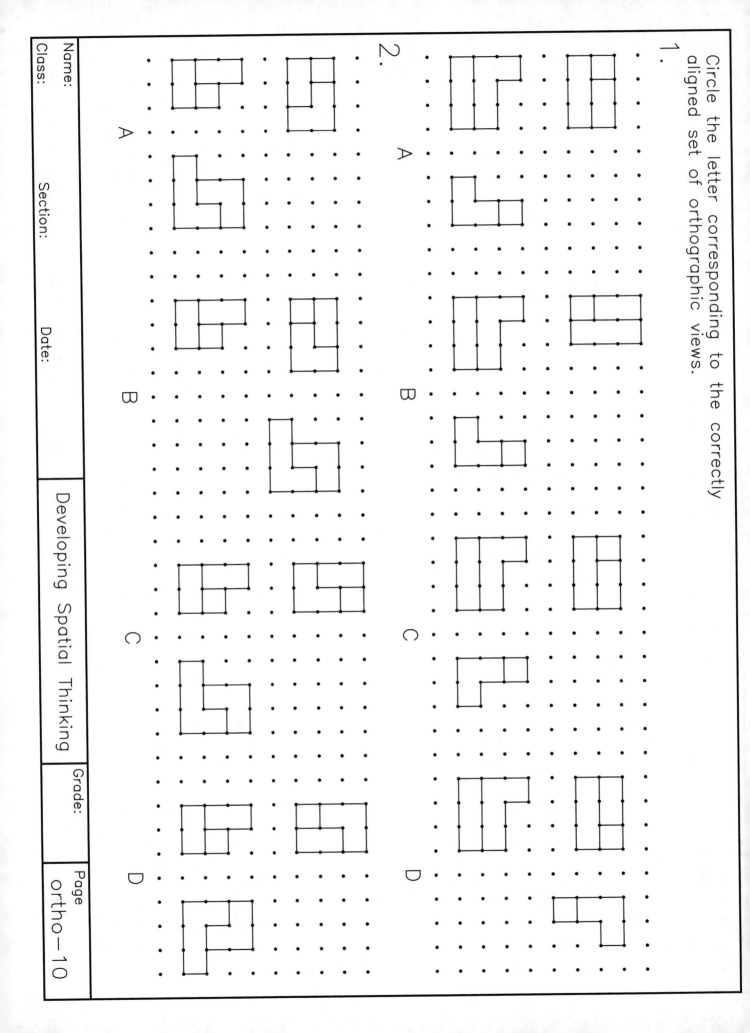

2.

A

B

C

D

Name:
Class:
Section:
Date:
Developing Spatial Thinking
Grade:
Page
ortho-10

Circle the letter corresponding to the correctly aligned set of orthographic views.

1.

2.

A   B   C   D

A   B

C   D

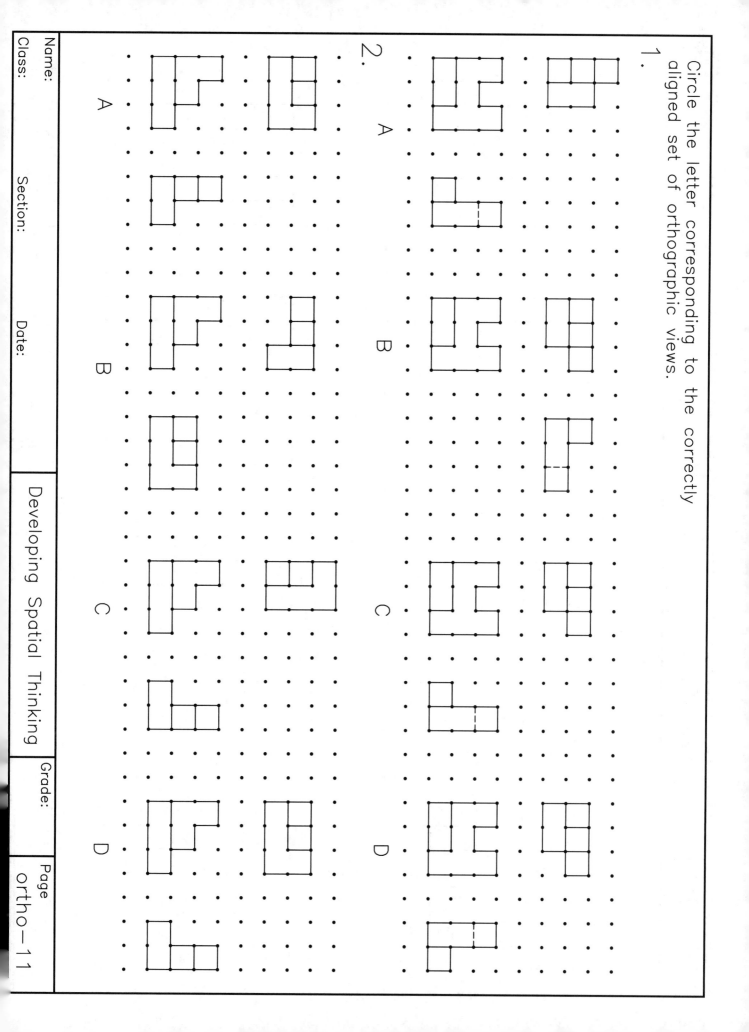

Developing Spatial Thinking

Circle the letter corresponding to the correctly aligned set of orthographic views.

1.

2.

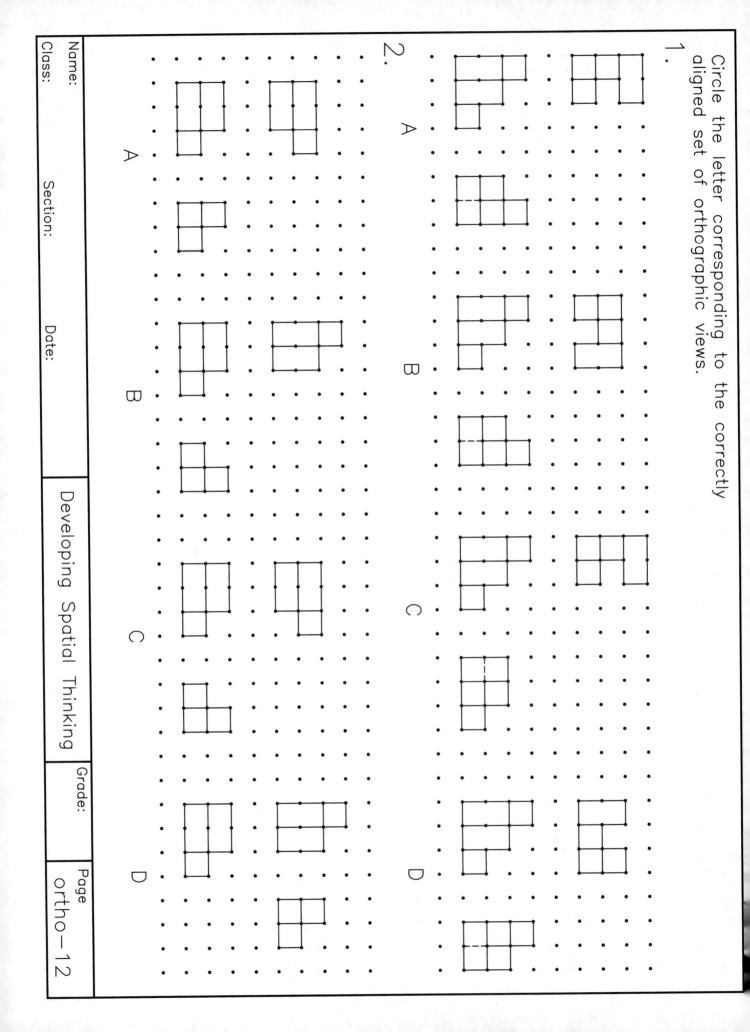

Name:
Class:

Section:

Date:

Developing Spatial Thinking

Grade:

Page
ortho-12

Circle the letter corresponding to the correctly aligned set of orthographic views.

1.

A

B

C

D

2.

A

B

C

D

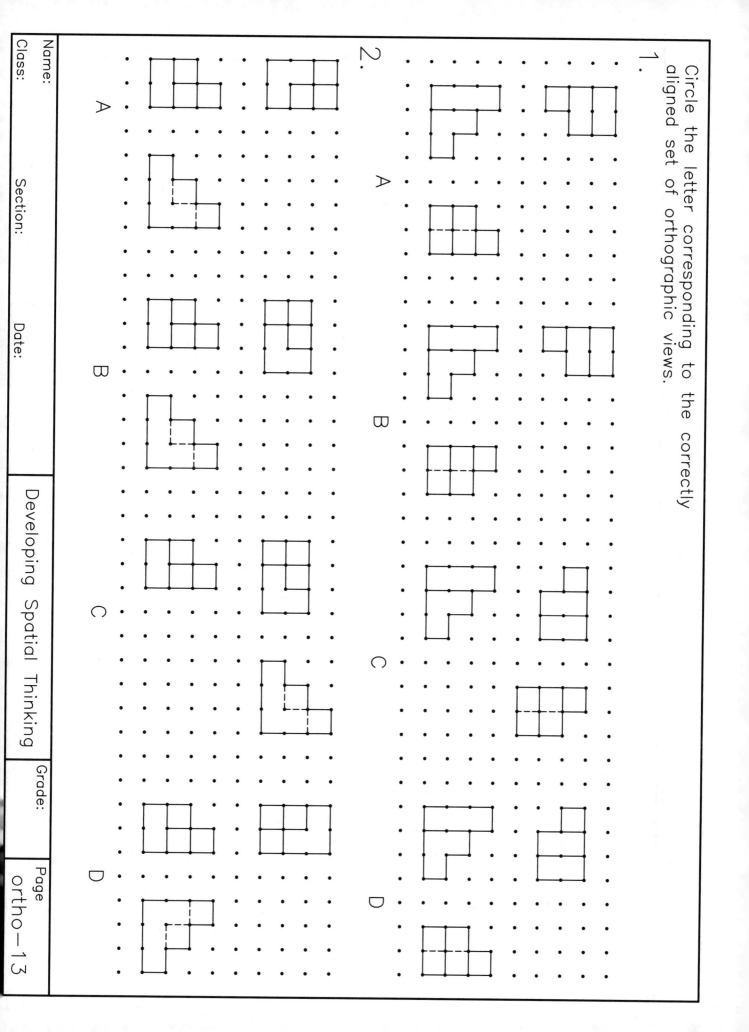

Circle the letter corresponding to the correctly aligned set of orthographic views.

1.

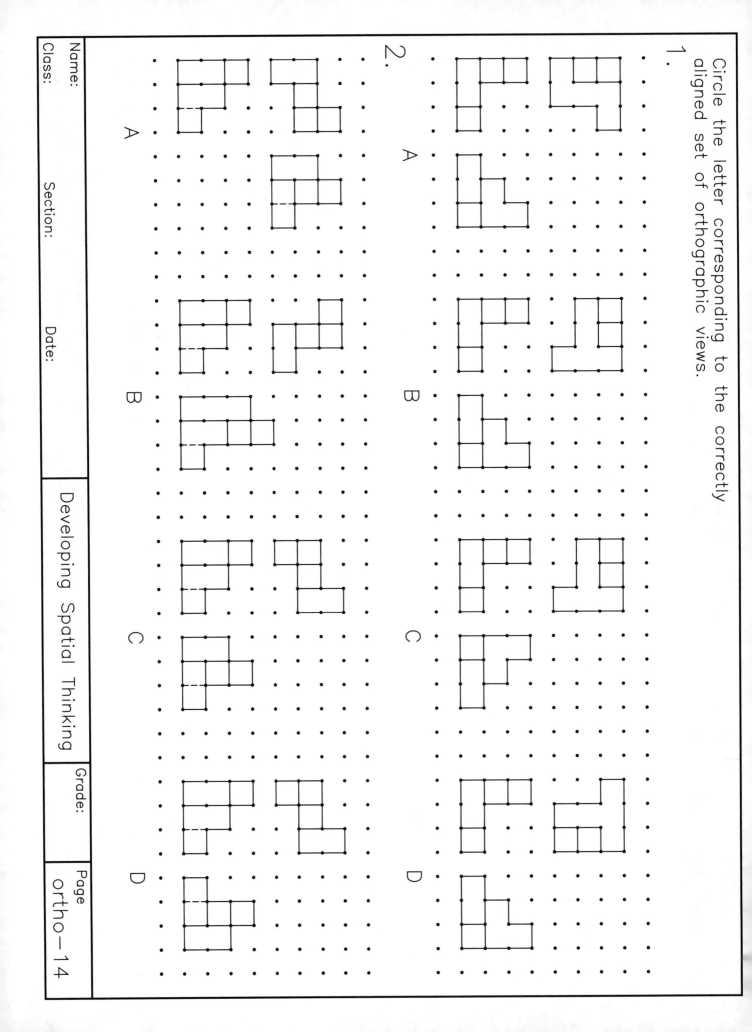

A

B

C

D

2.

A

B

C

D

For the objects shown in isometric below, sketch the top, front, and right side views in the space provided. Make sure that your views are properly aligned.

1.

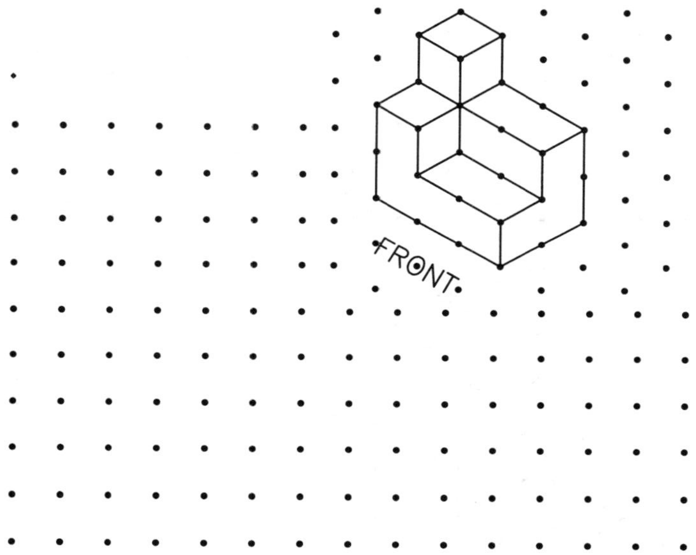

FRONT.

2.

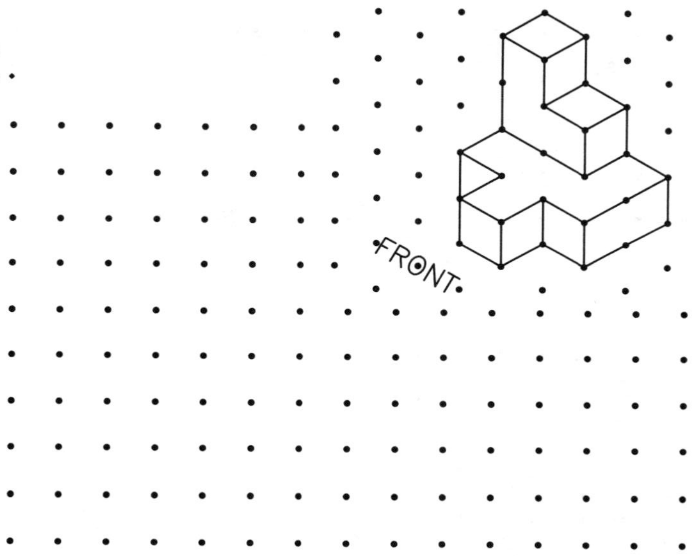

FRONT.

3.

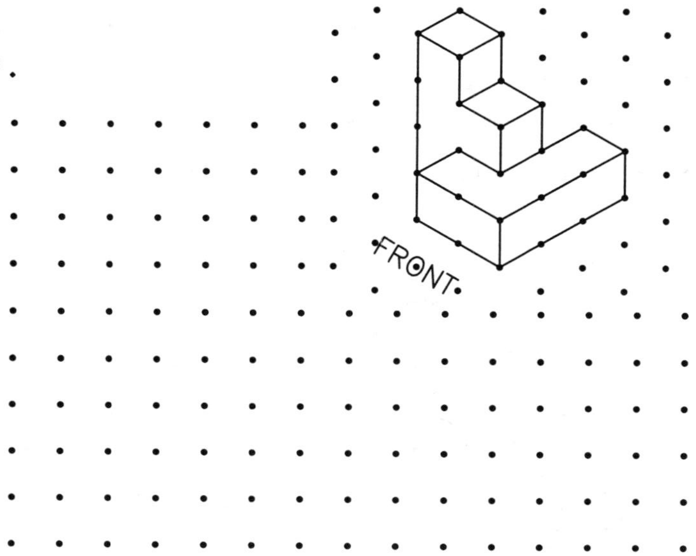

FRONT.

4.

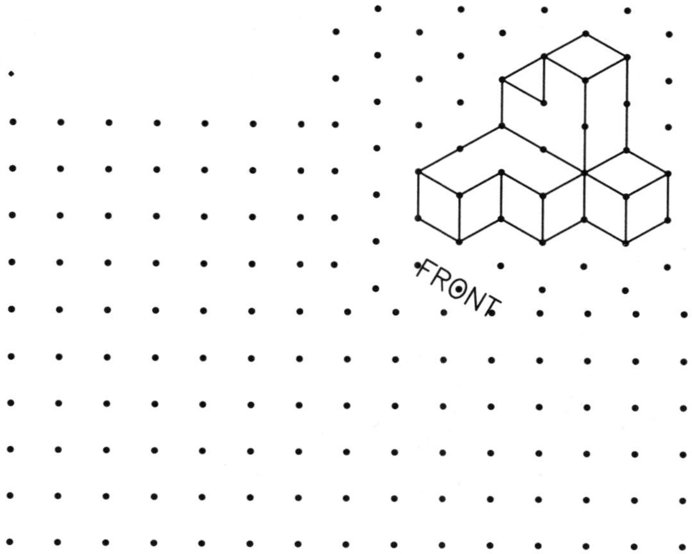

FRONT.

For the objects shown in isometric below, sketch the top, front, and right side views in the space provided. Make sure that your views are properly aligned.

1.

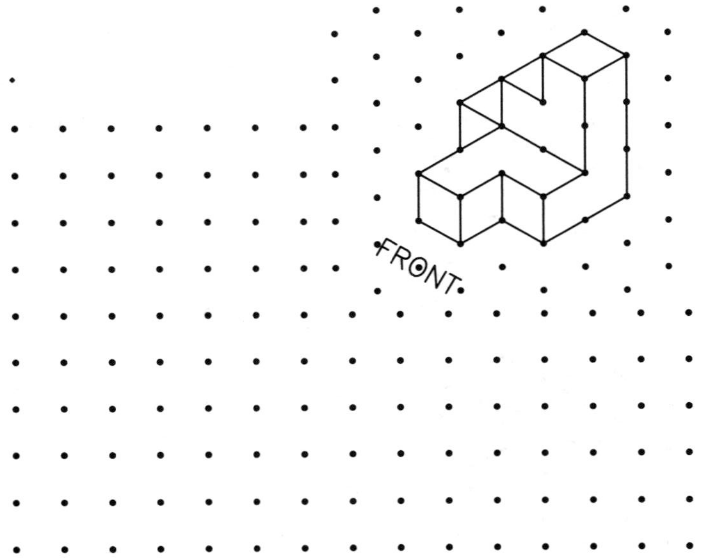

FRONT

2.

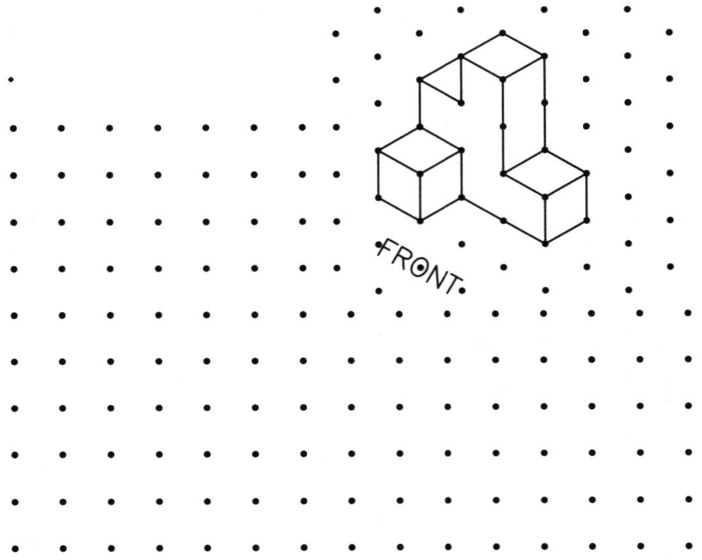

FRONT

3.

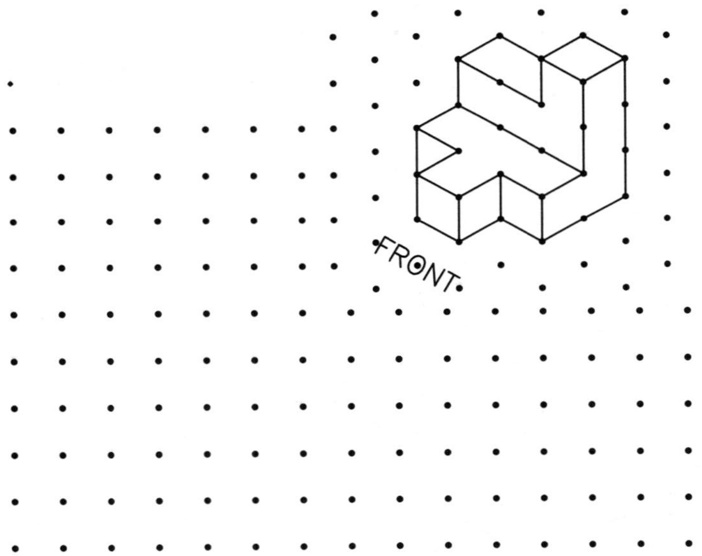

FRONT

4.

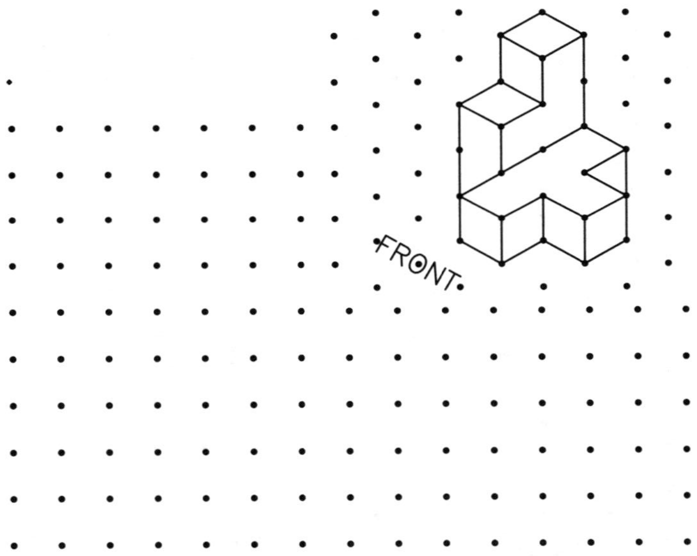

FRONT

For the objects shown in isometric below, sketch the top, front, and right side views in the space provided. Make sure that your views are properly aligned.

1.

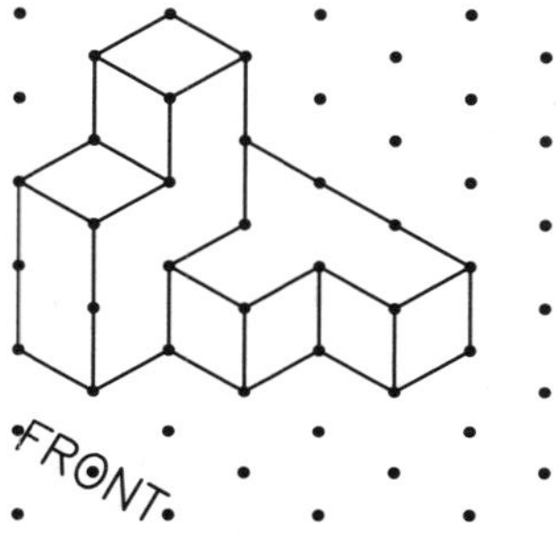

FRONT

2.

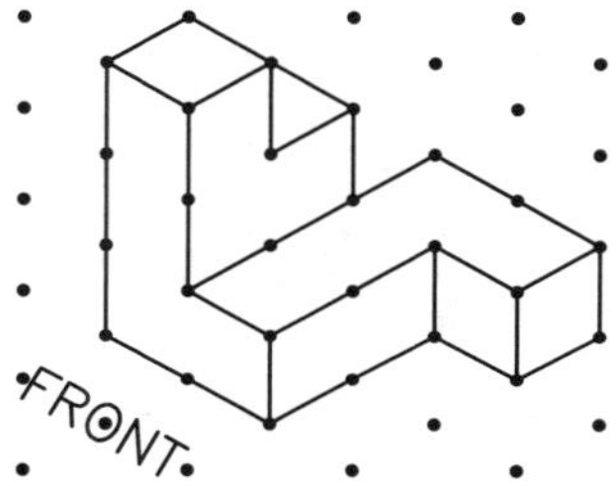

FRONT

3.

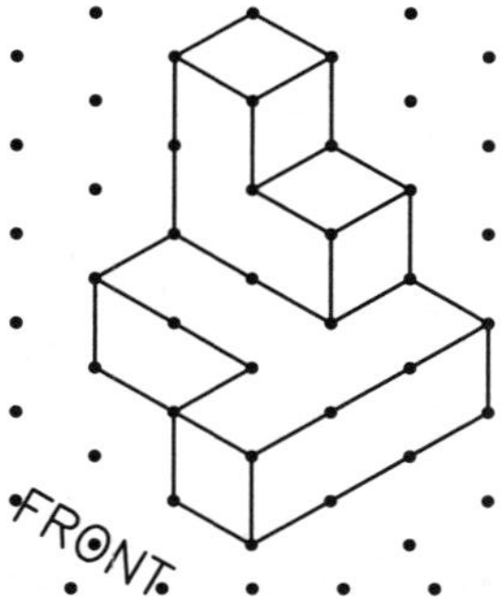

FRONT

4.

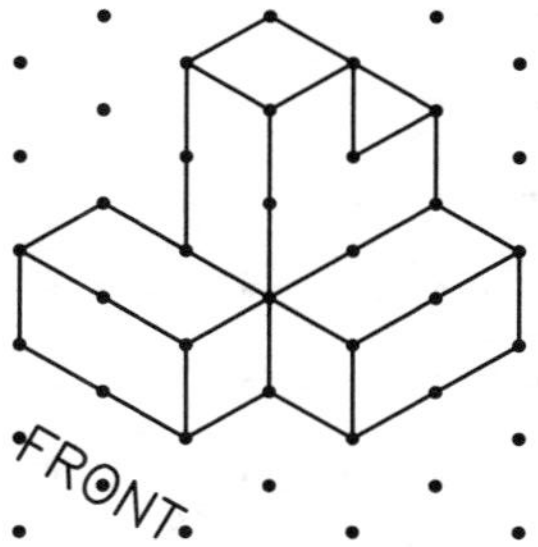

FRONT

Name:

Class:

Section:

Date:

Developing Spatial Thinking

Grade:

Page
ortho—17

For the objects shown in isometric below, sketch the top, front, and right side views in the space provided. Make sure that your views are properly aligned.

1.

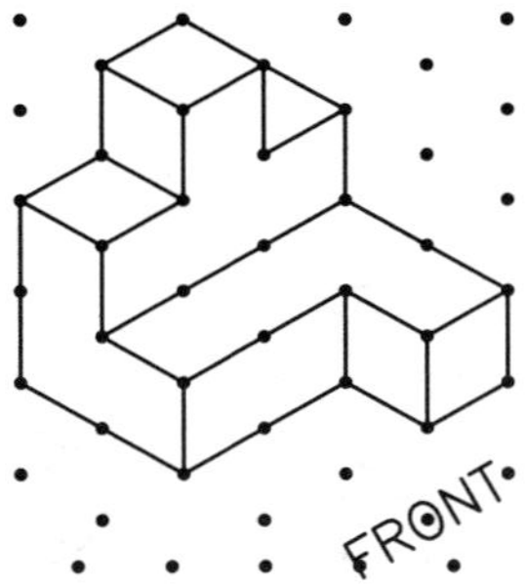

FRONT

2.

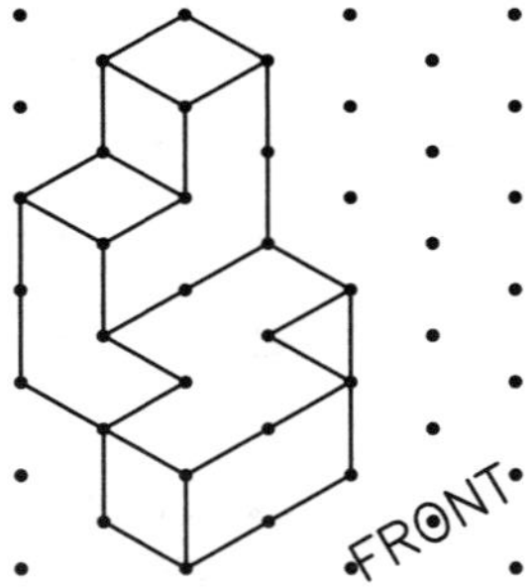

FRONT

3.

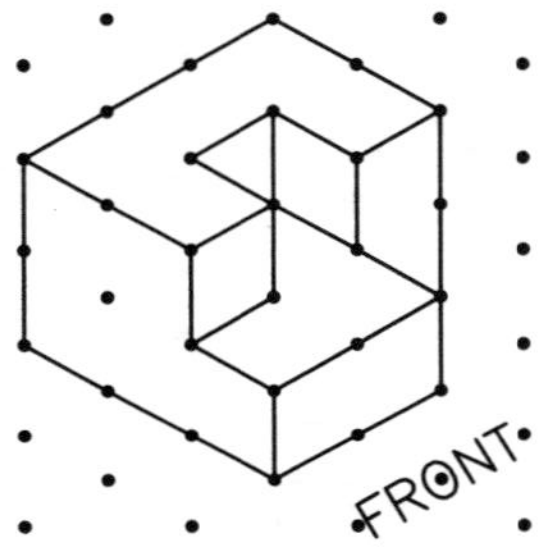

FRONT

4.

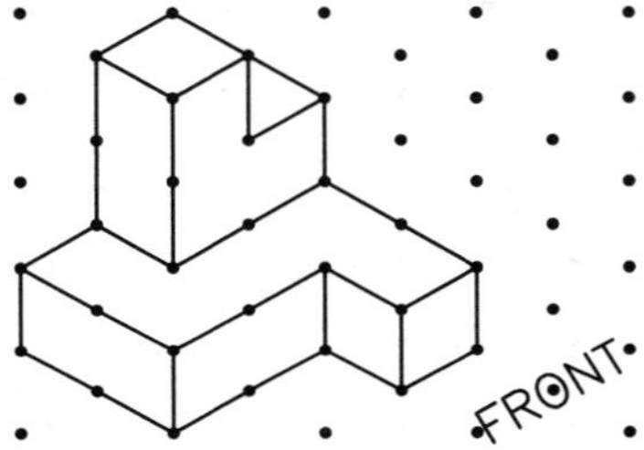

FRONT

For the objects shown in isometric below, sketch the top, front, and right side views in the space provided. Make sure that your views are properly aligned.

1.

2.

3.

FRONT

4.

FRONT

FRONT

FRONT

Name:

Class:

Section:

Date:

Developing Spatial Thinking

Grade:

Page
ortho—19

For the objects shown in orthographic projection below, construct an isometric view in the space provided. Use the box method to assist you if necessary.

1.

2.

3.

4.

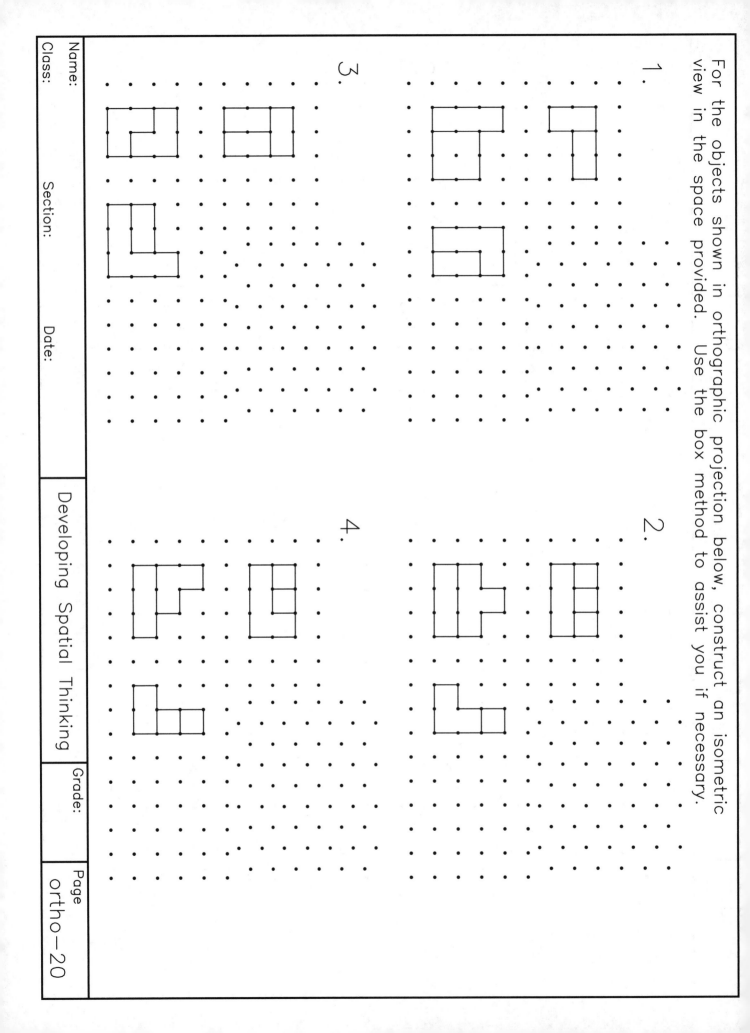

Name:

Class:

Section:

Date:

Developing Spatial Thinking

Grade:

Page
ortho—20

For the objects shown in orthographic projection below, construct an isometric view in the space provided. Use the box method to assist you if necessary.

1.

2.

3.

4.

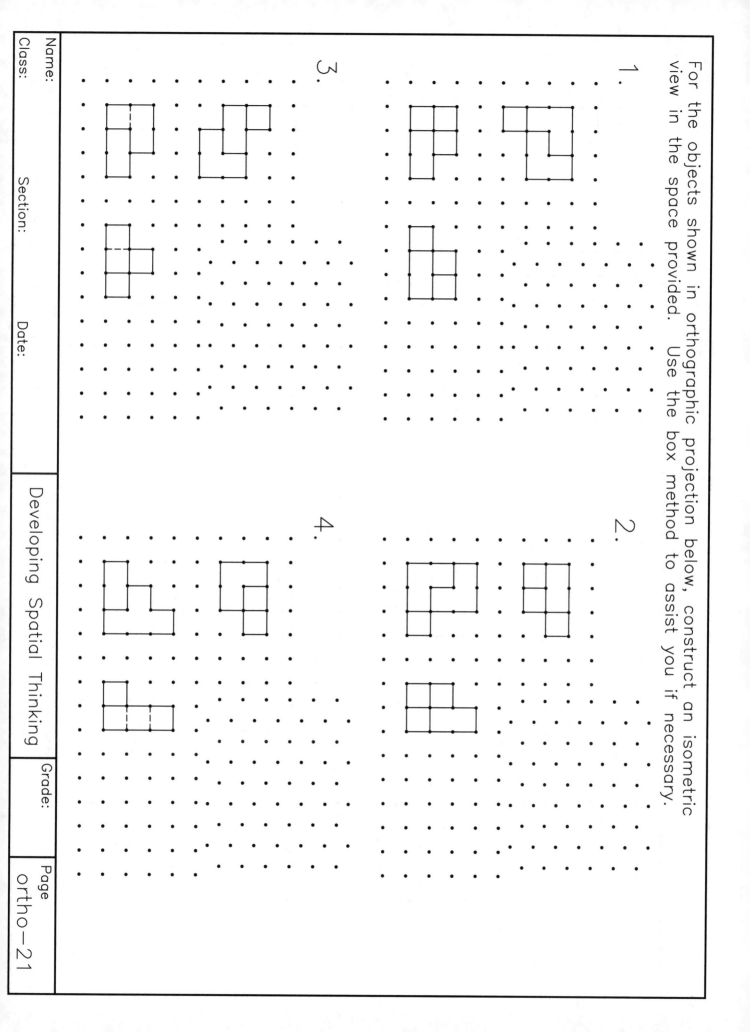

Name:
Class:

Section:

Date:

Developing Spatial Thinking

Grade:

Page
ortho-21

For the objects shown in orthographic projection below, construct an isometric view in the space provided. Use the box method to assist you if necessary.

1.

2.

3.

4.

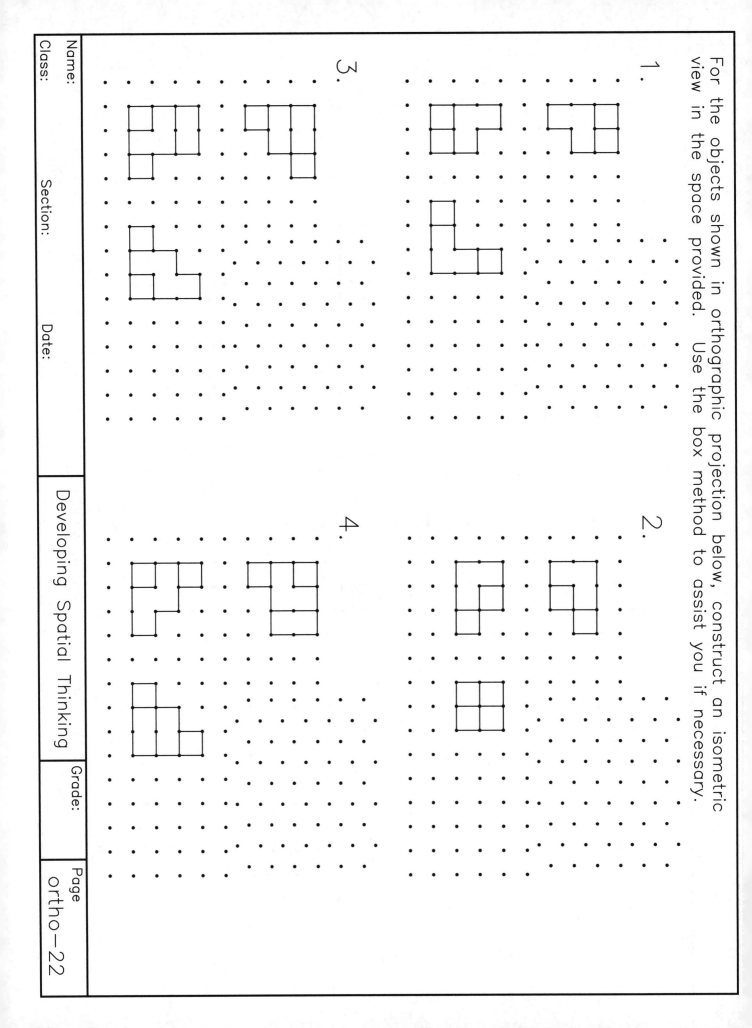

Name:
Class:

Section:

Date:

Developing Spatial Thinking

Grade:

Page
ortho-22

For the objects shown in orthographic projection below, construct an isometric view in the space provided. Use the box method to assist you if necessary.

1.

2.

3.

4.

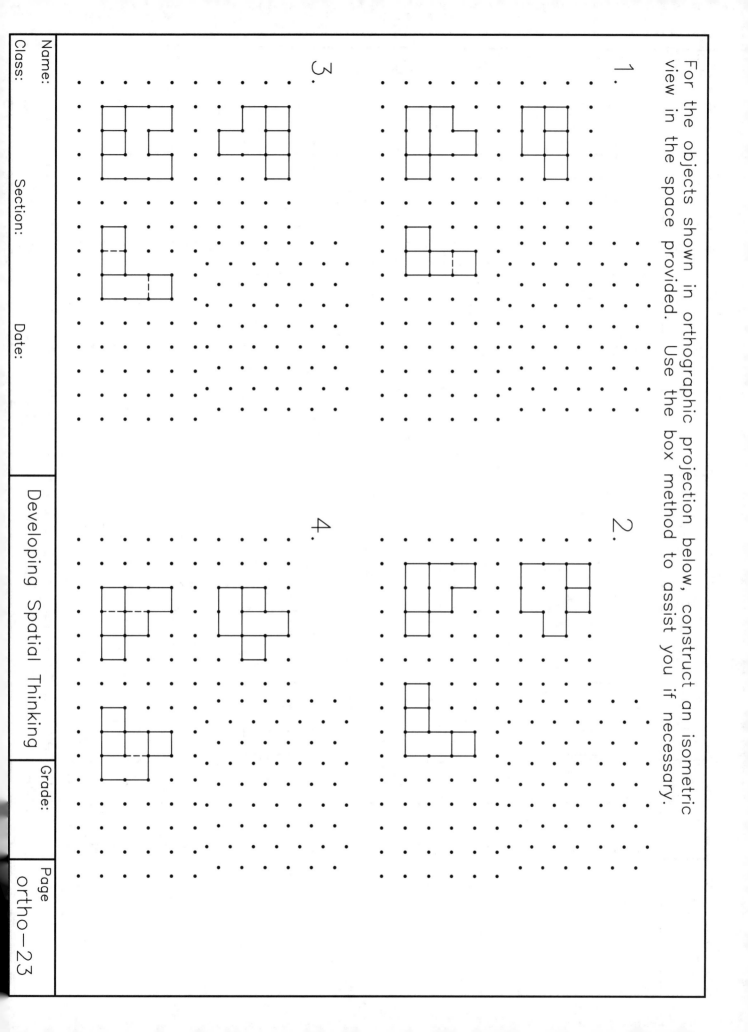

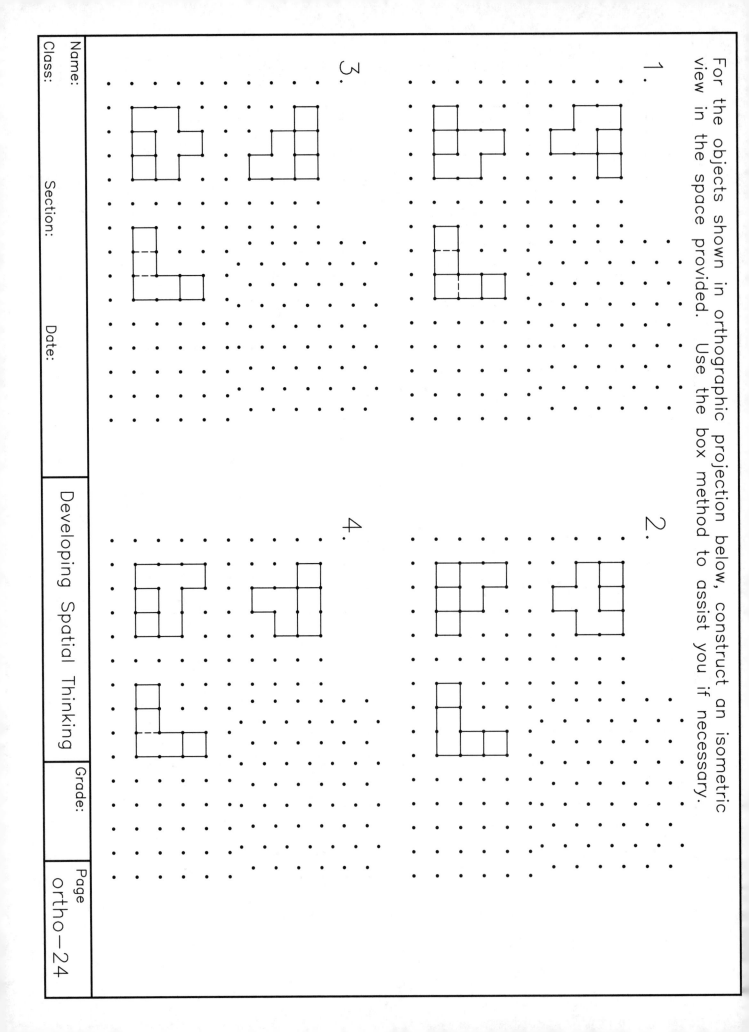

For the objects shown in orthographic projection below, construct an isometric view in the space provided. Use the box method to assist you if necessary.

1.

2.

3.

4.

Name:
Class:

Section:

Date:

Developing Spatial Thinking

Grade:

Page
ortho-24

Inclined and Curved Surfaces

Orthographic drawings are used to show the top, front, and side views of an object. In creating orthographic views, recall that you first imagine that the object is surrounded by a glass cube. The edges and surfaces are projected onto the panes of glass and the cube is unfolded so that the panes of glass all lie in the same plane.

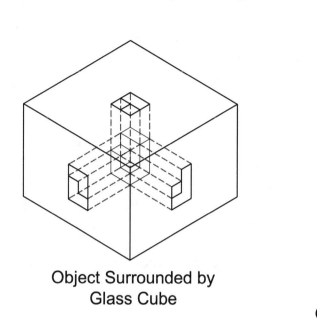

Object Surrounded by
Glass Cube

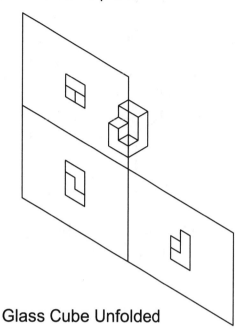

Glass Cube Unfolded

Normal surfaces are defined as being parallel to either the top, front, or side views. A normal surface is seen as a surface in the view to which it is parallel and is seen as an edge in the other views

Many objects contain surfaces that are angled with respect to the top, front, or side views. These surfaces are called inclined surfaces.

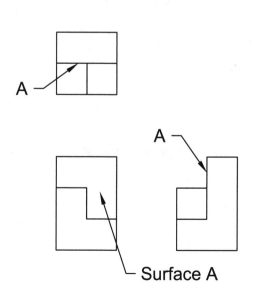

Surface A

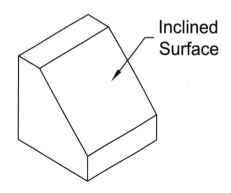

Inclined
Surface

Orthographic drawings of objects with inclined surfaces are constructed in the same way as the drawings for objects with only normal surfaces. Imagine that the object is surrounded by a glass cube, project the edges and surfaces to the panes of glass, and unfold the glass cube so its panes are in one plane.

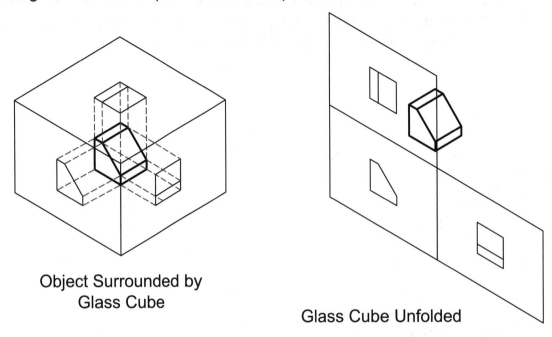

Object Surrounded by
Glass Cube

Glass Cube Unfolded

When projecting inclined surfaces they appear *foreshortened* in the orthographic views where they appear as surfaces. This means that they look *smaller* than their actual size.

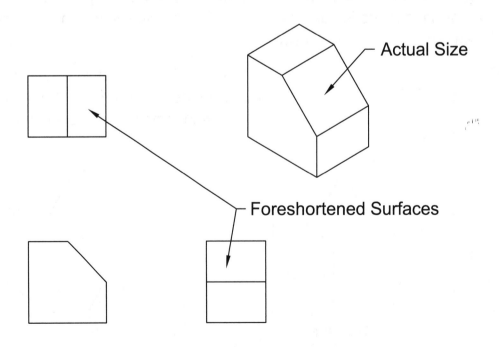

Actual Size

Foreshortened Surfaces

Recall that normal surfaces appear as an edge in two views and as an area in one view. Inclined surfaces appear as an edge in one view and as an area in two views. The edge view of the inclined surfaces is "angled."

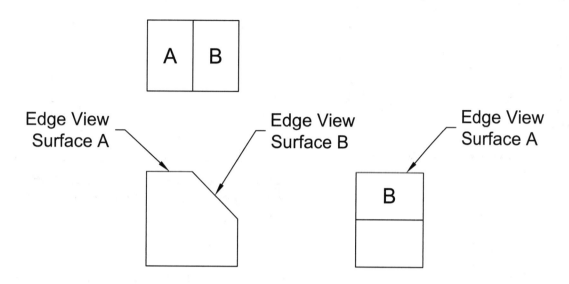

An inclined surface maintains its basic shape from view to view.

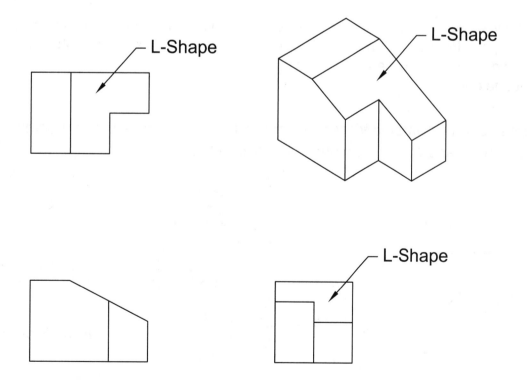

All of the rules for orthographic projection of normal surfaces, still apply to inclined surfaces.

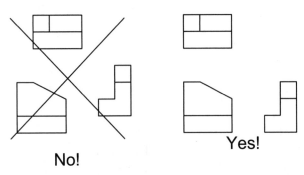

No!   Yes!

The orthographic views of an object should be aligned with one another.

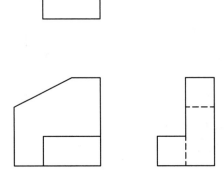

Edges of an object that are hidden from a given viewpoint are shown as dashed lines in that view.

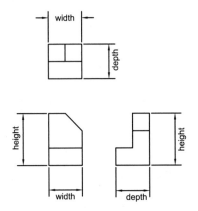

The top view shows the width and depth of an object, the front view shows the height and width, and the side view shows the height and depth.

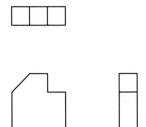

When a hidden line coincides with a solid line on a view, show only the solid line.

To construct an isometric view of an inclined surface, locate the end points of each inclined edge and draw a straight line between them.

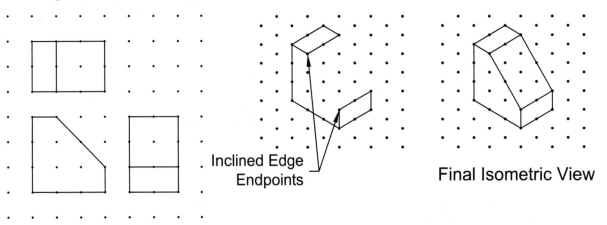

Orthographic Views

Inclined Edge Endpoints

Final Isometric View

When constructing an isometric view of an inclined surface, choose an orientation that best shows the inclined surface.

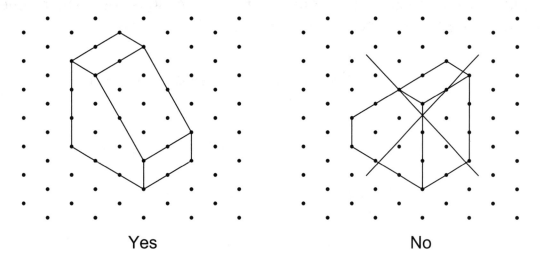

Yes   No

Single curved surfaces are generated by revolving a line about an axis.

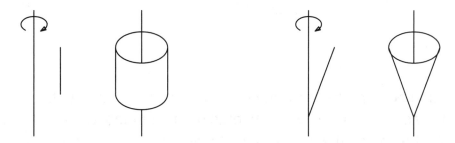

For an orthographic drawing of an object such as a cylinder, you can again imagine the object surrounded by a glass box. In this case, there are no sharp edges on the object, so the visible extents of the object are projected onto the panes of glass.

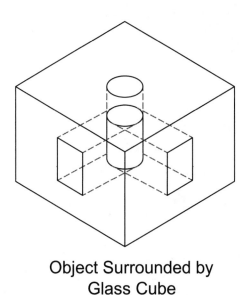

Object Surrounded by
Glass Cube

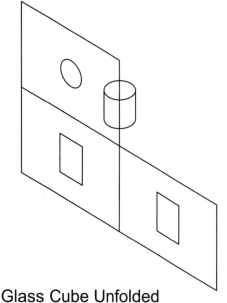

Glass Cube Unfolded

A cylinder will appear as a circle in one view and as rectangles in the other two views. To make sure that others understand that the rectangle is actually an area view of a curved surface, center lines are included that consist of two long dashes separated by one short dash. In the end view of the cylinder, crossing centerlines are included.

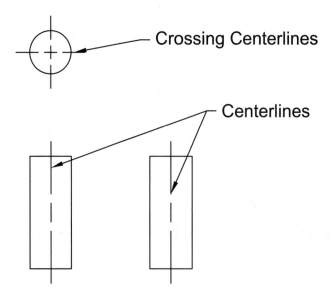

Crossing Centerlines

Centerlines

The most common occurrence of cylindrical surfaces are for holes that go through objects. In this case, you still show the extents of the surface, but since the curved surface is <u>inside</u> the object, these lines are shown as hidden (dashed) lines.

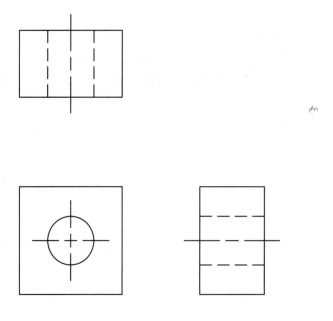

For the object shown in orthographic projection on the left, circle the letter of the correct corresponding isometric view.

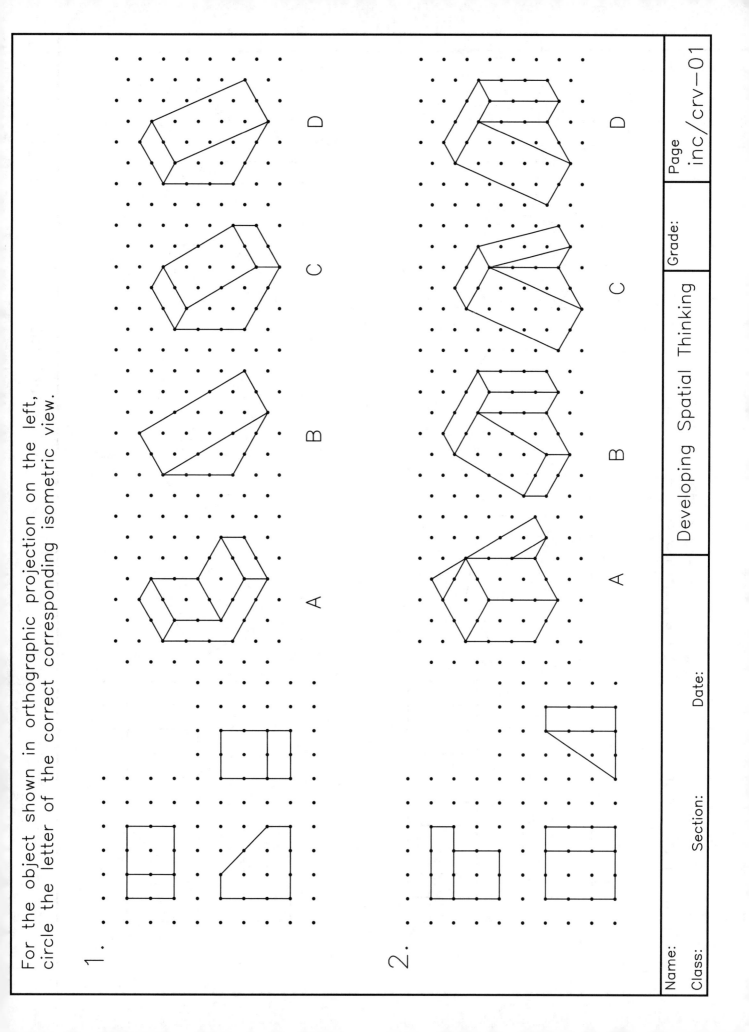

1.

A   B   C   D

2.

A   B   C   D

Name:

Class:

Section:

Date:

Developing Spatial Thinking

Grade:

Page
inc/crv−01

For the object shown in orthographic projection on the left, circle the letter of the correct corresponding isometric view.

1.

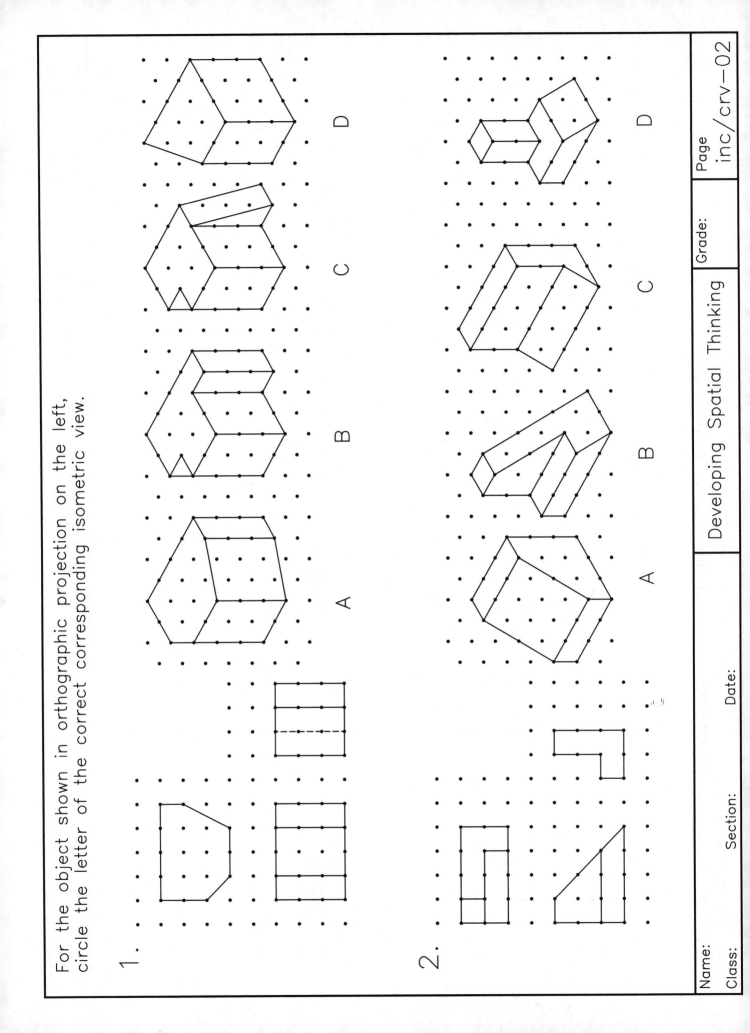

A B C D

2.

A B C D

Name:
Class: Section: Date:

Developing Spatial Thinking   Grade: Page
inc/crv-02

For the object shown in orthographic projection on the left, circle the letter of the correct corresponding isometric view.

1.

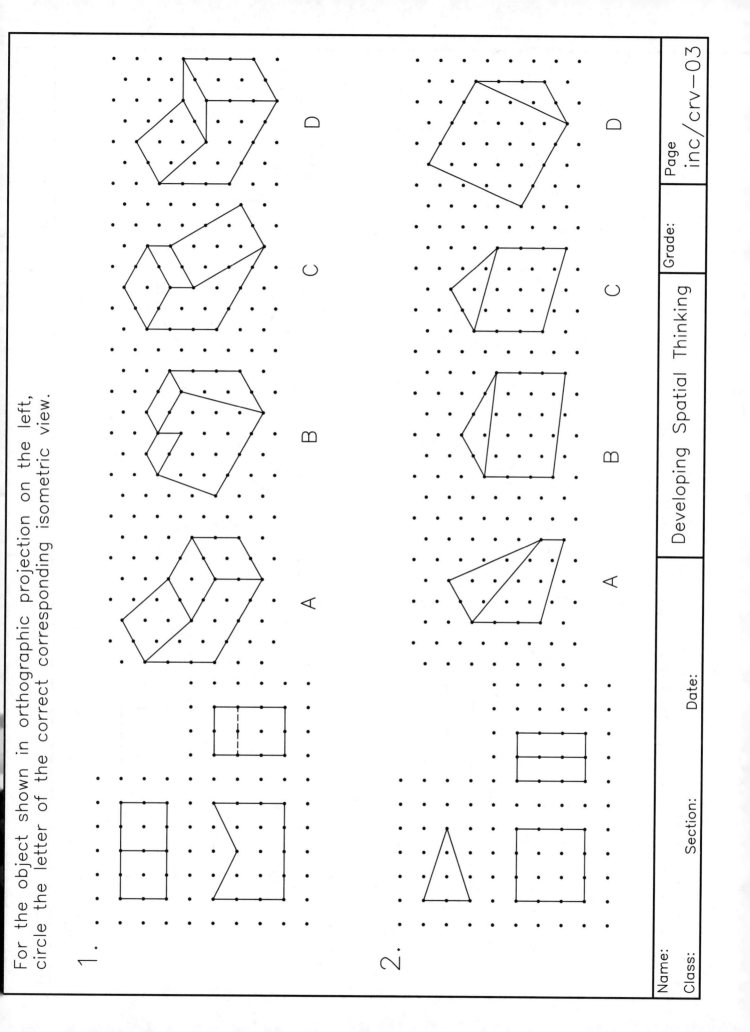

A   B   C   D

2.

A   B   C   D

Name:

Class:

Section: Date:

Developing Spatial Thinking

Grade:

Page
inc/crv-03

For the object shown in orthographic projection on the left,
circle the letter of the correct corresponding isometric view.

1.

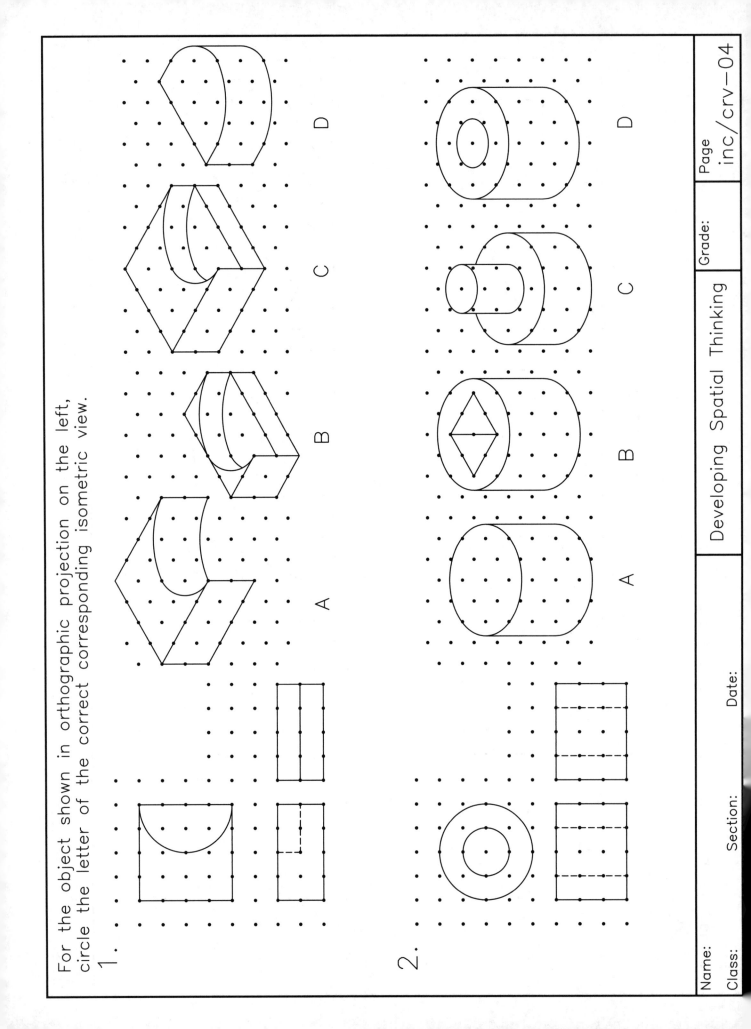

A   B   C   D

2.

A   B   C   D

For the object shown in orthographic projection on the left, circle the letter of the correct corresponding isometric view.

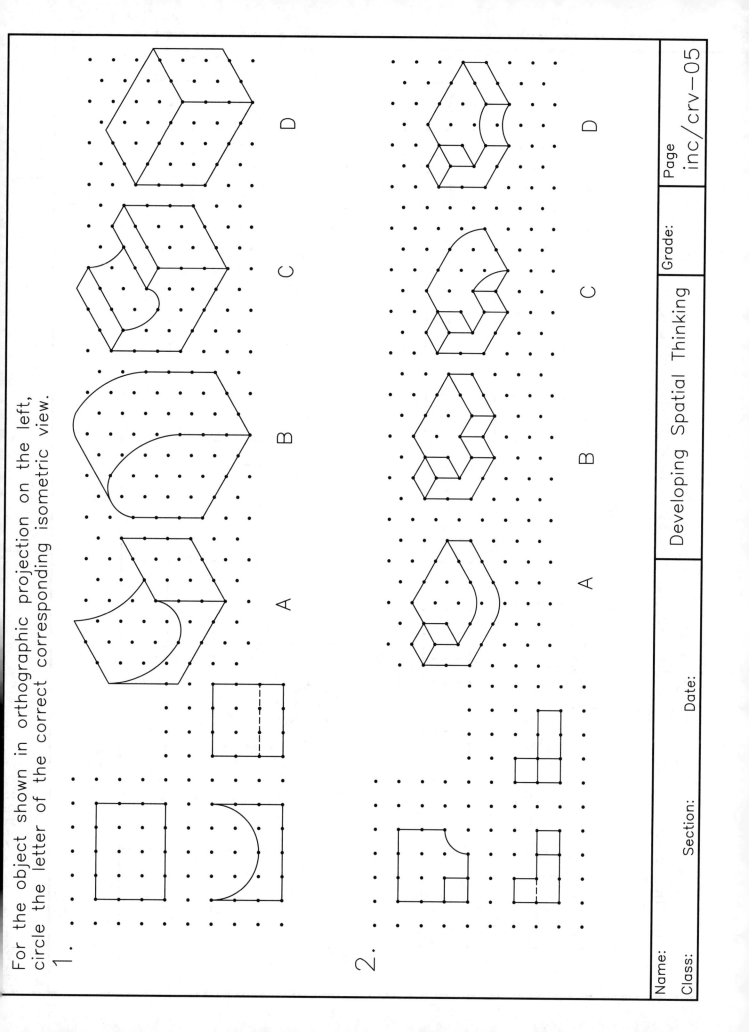

1.

A   B   C   D

2.

A   B   C   D

Name:

Class: Section: Date:

Developing Spatial Thinking

Grade:

Page
inc/crv—05

An isometric view of an object is shown below along with its top and front views. Circle the letter corresponding to the correct side view from the choices given.

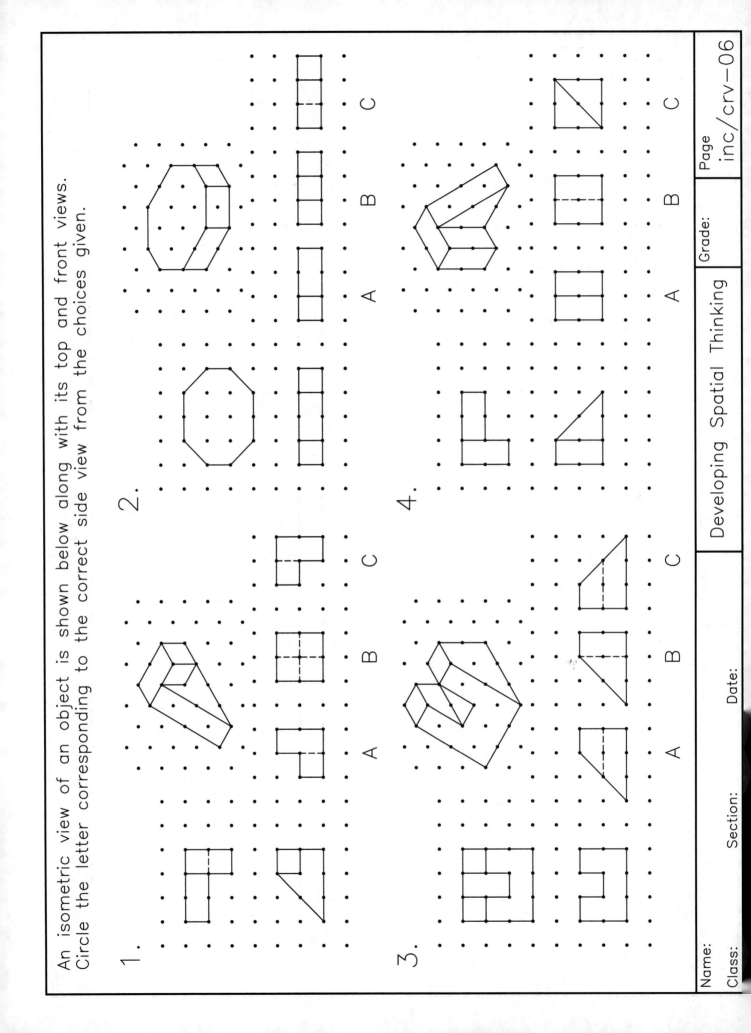

1.

2.

3.

4.

A   B   C

Developing Spatial Thinking

Page
inc/crv—06

Name:

Class:

Section:

Date:

Grade:

An isometric view of an object is shown below along with its top and front views. Circle the letter corresponding to the correct side view from the choices given.

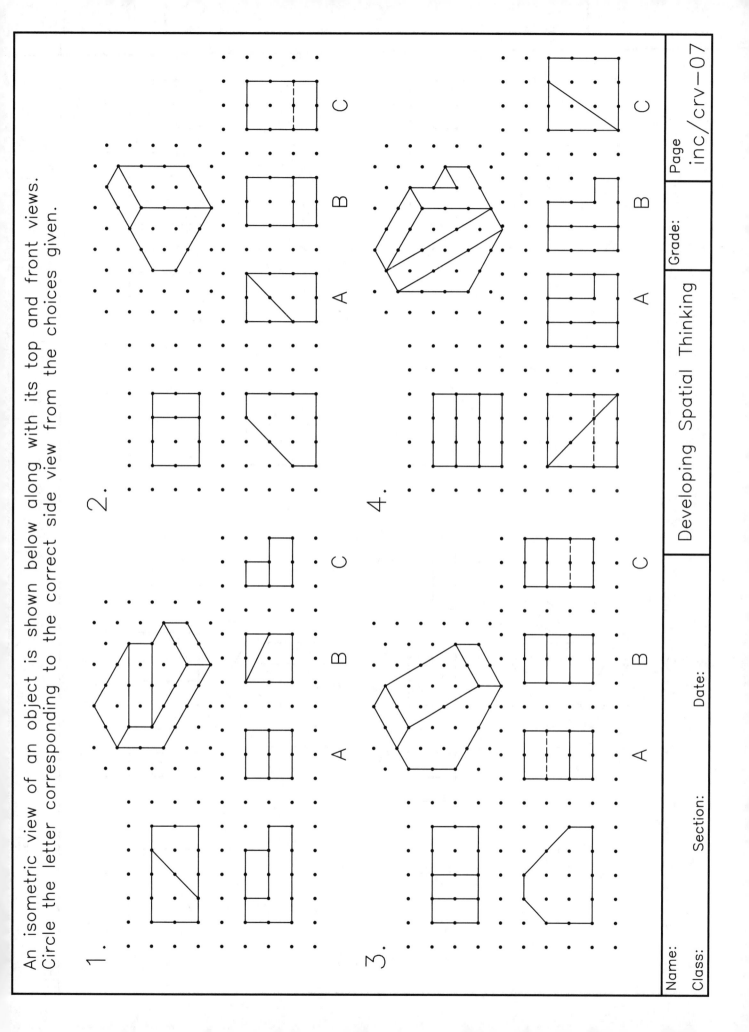

1.

2.

3.

4.

A   B   C

Developing Spatial Thinking

Name:

Class:

Section:

Date:

Grade:

An isometric view of an object is shown below along with its top and front views. Circle the letter corresponding to the correct side view from the choices given.

1.

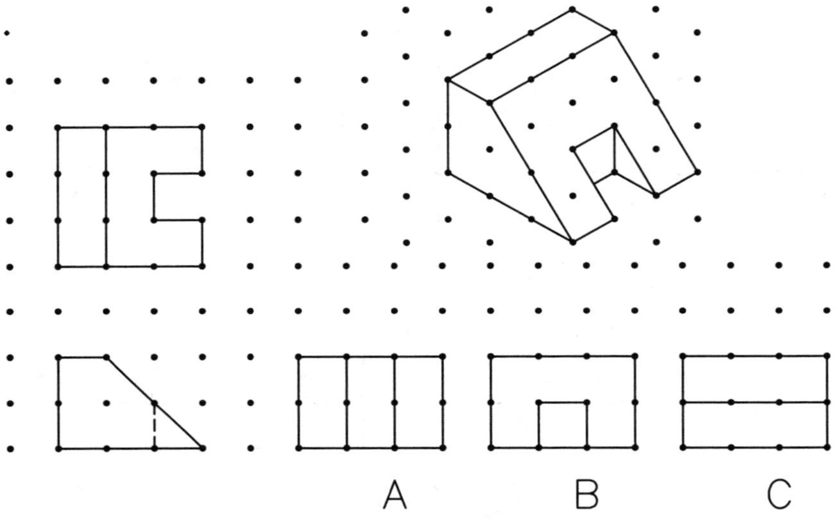

2.

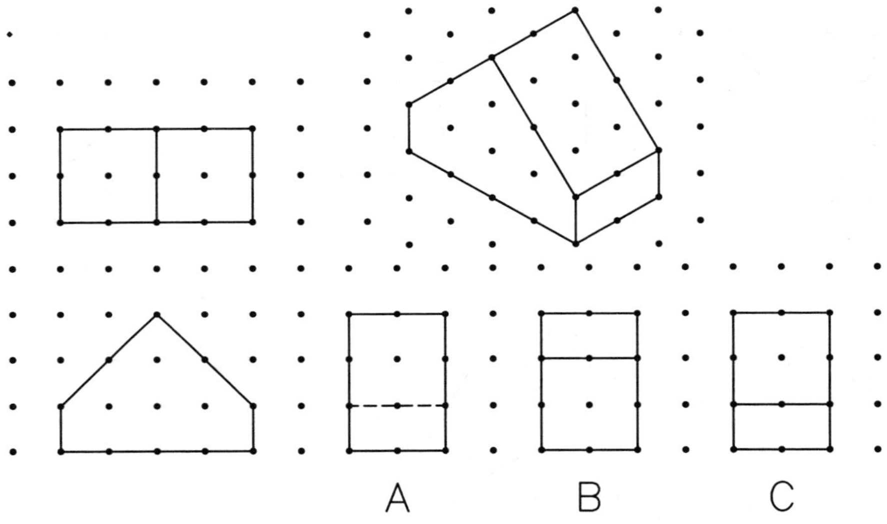

3.

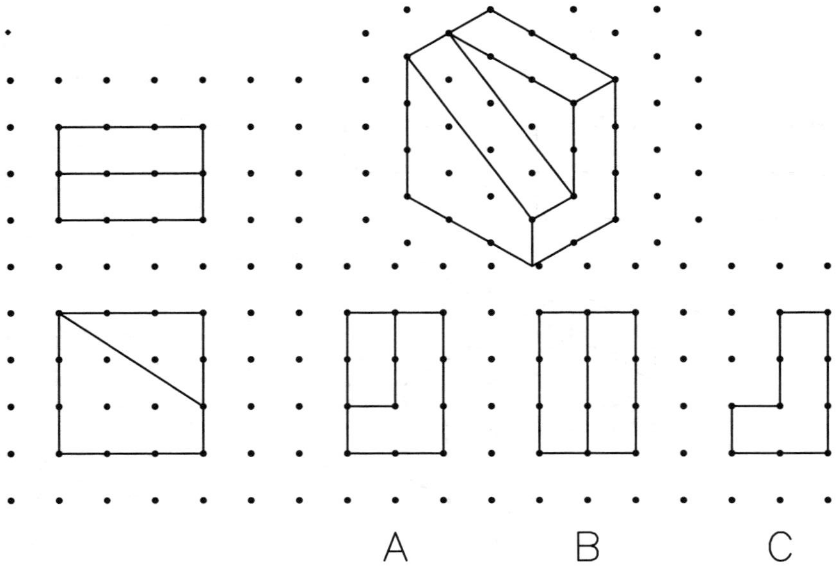

4.

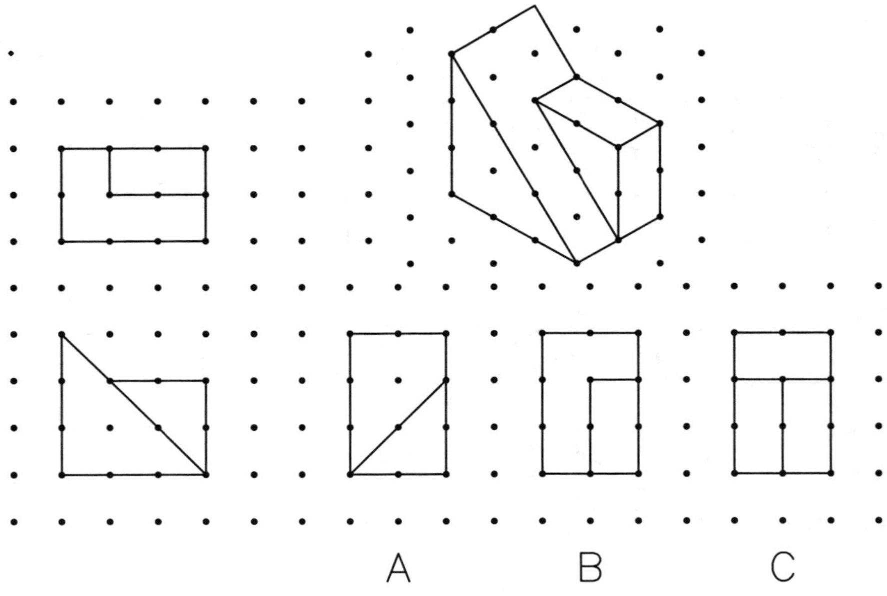

A   B   C

Name:

Class:

Section:

Date:

Developing Spatial Thinking

Grade:

Page
inc/crv-08

An isometric view of an object is shown below along with it's top and front views.
Circle the letter corresponding to the correct side view from the choices given

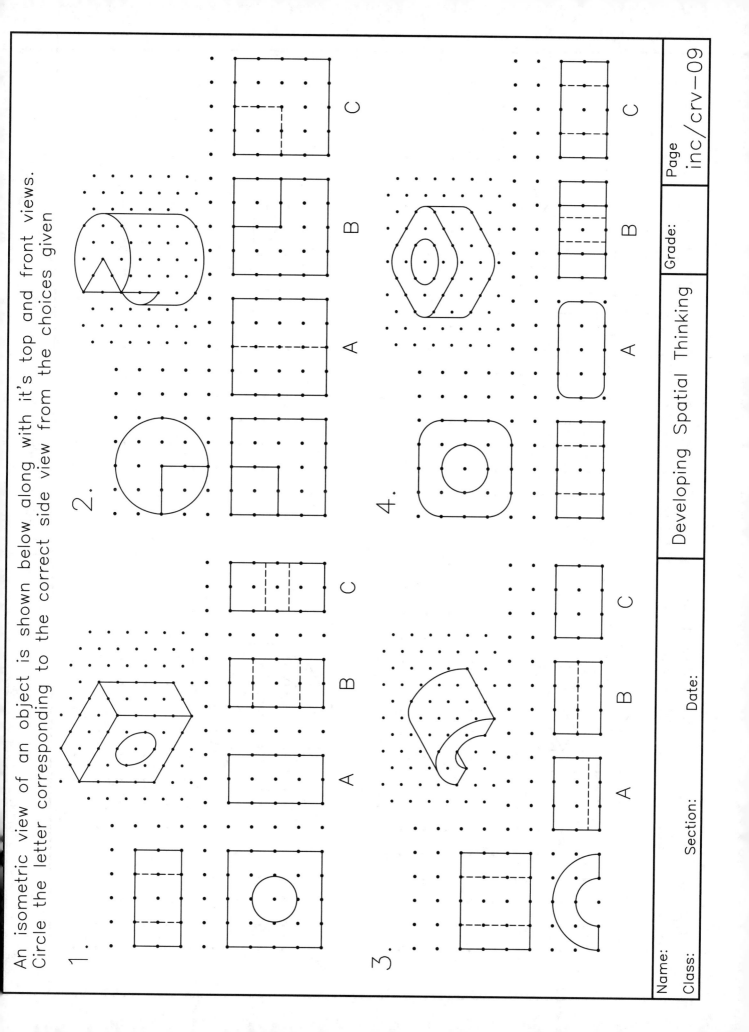

1.

2.

3.

4.

A   B   C

Developing Spatial Thinking

Name:

Class: Section: Date:

For the objects shown in isometric below, sketch the top, front, and right side views in the space provided. Make sure that your views are properly aligned.

1.

2.

FRONT

FRONT

3.

4.

FRONT

FRONT

Name:

Class:

Section:

Date:

Developing Spatial Thinking

Grade:

Page
inc/crv-10

For the objects shown in isometric below, sketch the top, front, and right side views in the space provided. Make sure that your views are properly aligned.

1.

FRONT

2.

FRONT

3.

FRONT

4.

FRONT

Name:

Class:

Section:

Date:

Developing Spatial Thinking

Grade:

Page
inc/crv-11

For the objects shown in isometric below, sketch the top, front, and right side views in the space provided. Make sure that your views are properly aligned.

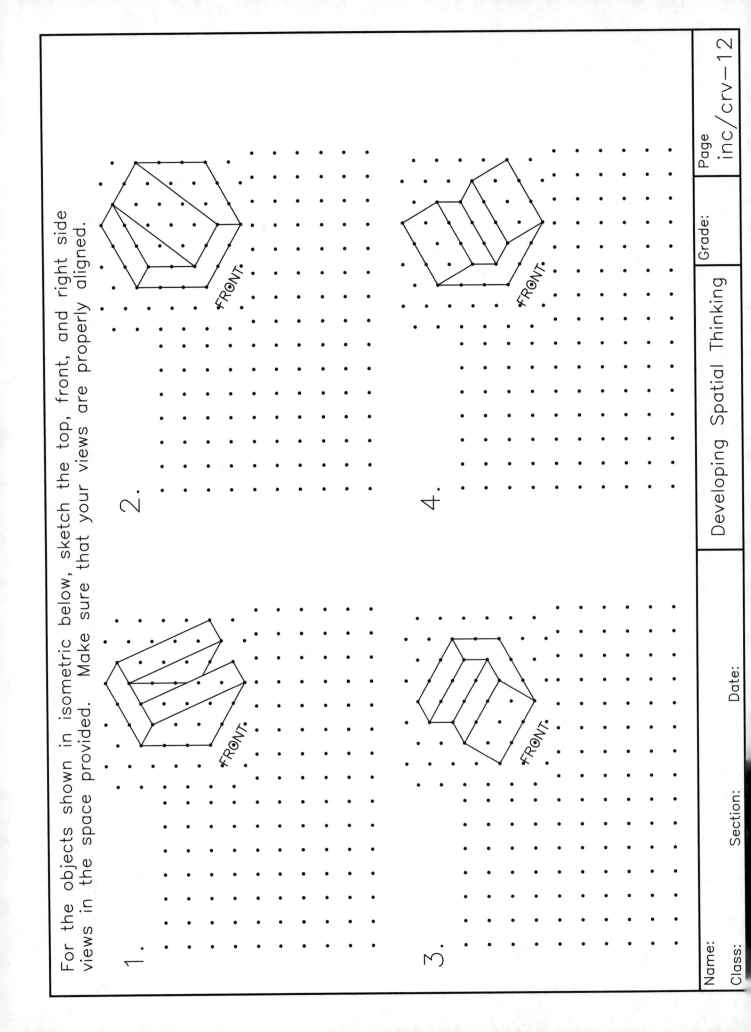

1.

2.   FRONT

3.   FRONT

4.   FRONT

Name:

Class: Section: Date:

Developing Spatial Thinking

Grade:

Page
inc/crv-12

For the objects shown in isometric below, sketch the top, front, and right side

1.

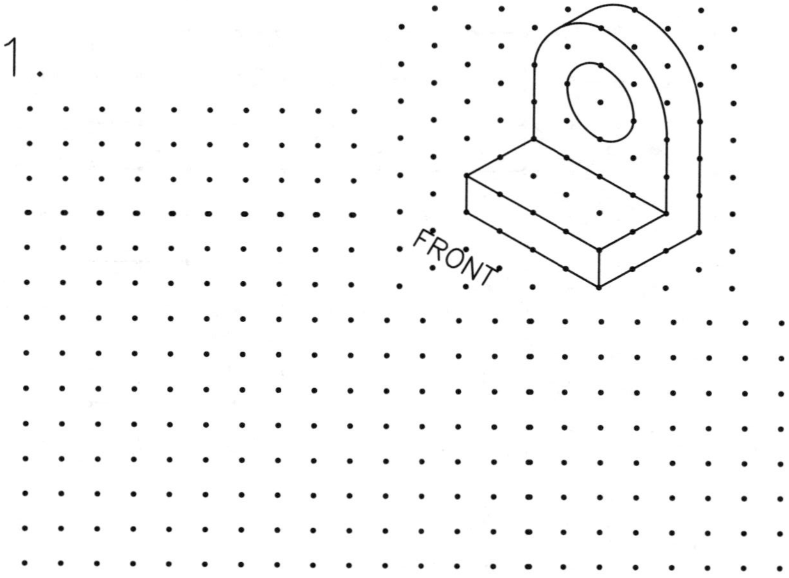

FRONT

2.

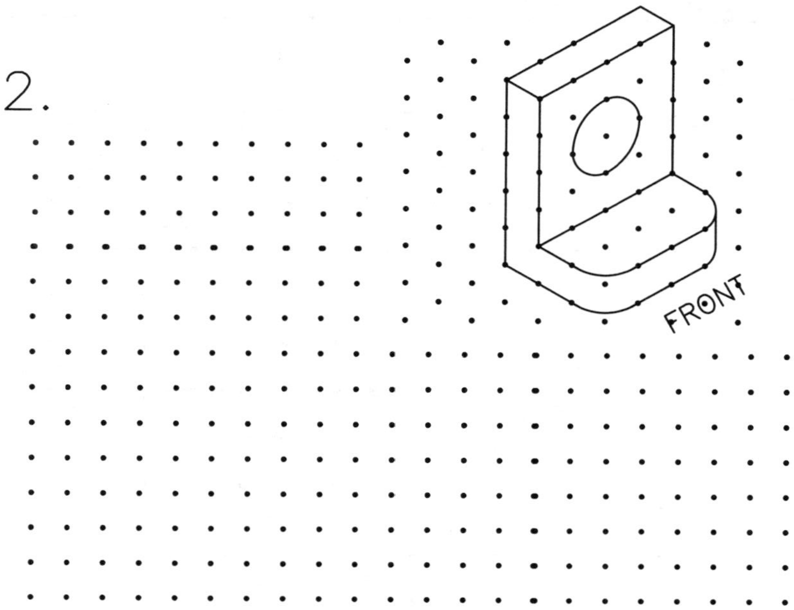

FRONT

3.

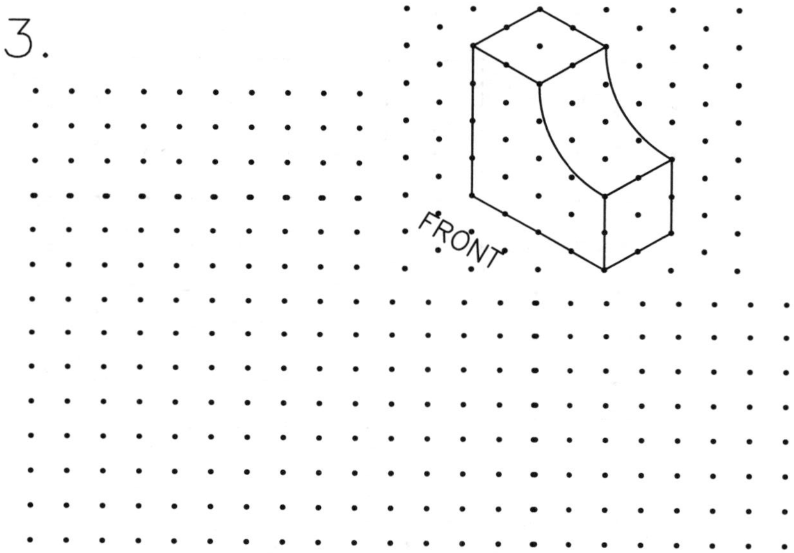

FRONT

4.

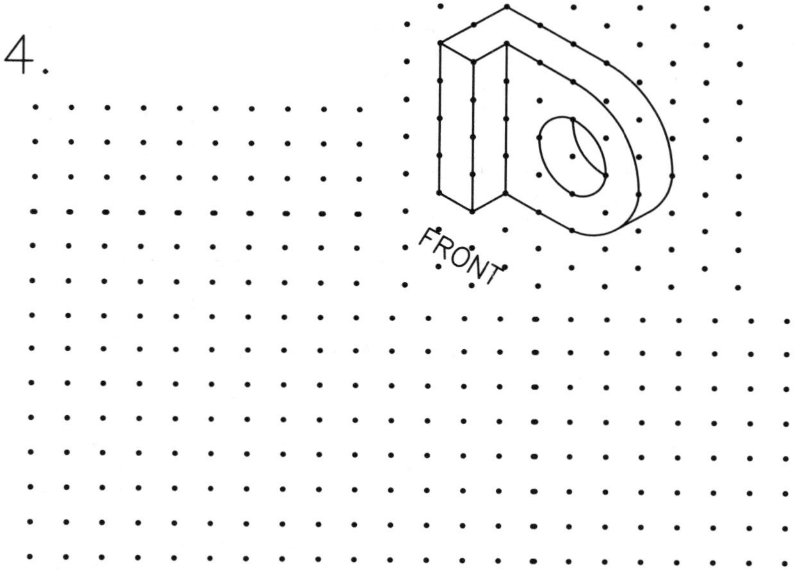

FRONT

Name:

Class:

Section:

Date:

Developing Spatial Thinking

Grade:

Page
inc/crv—13

For the objects shown in orthographic projection below, construct an isometric view in the space provided. Use the box method to assist you if necessary.

1.

2.

3.

4.

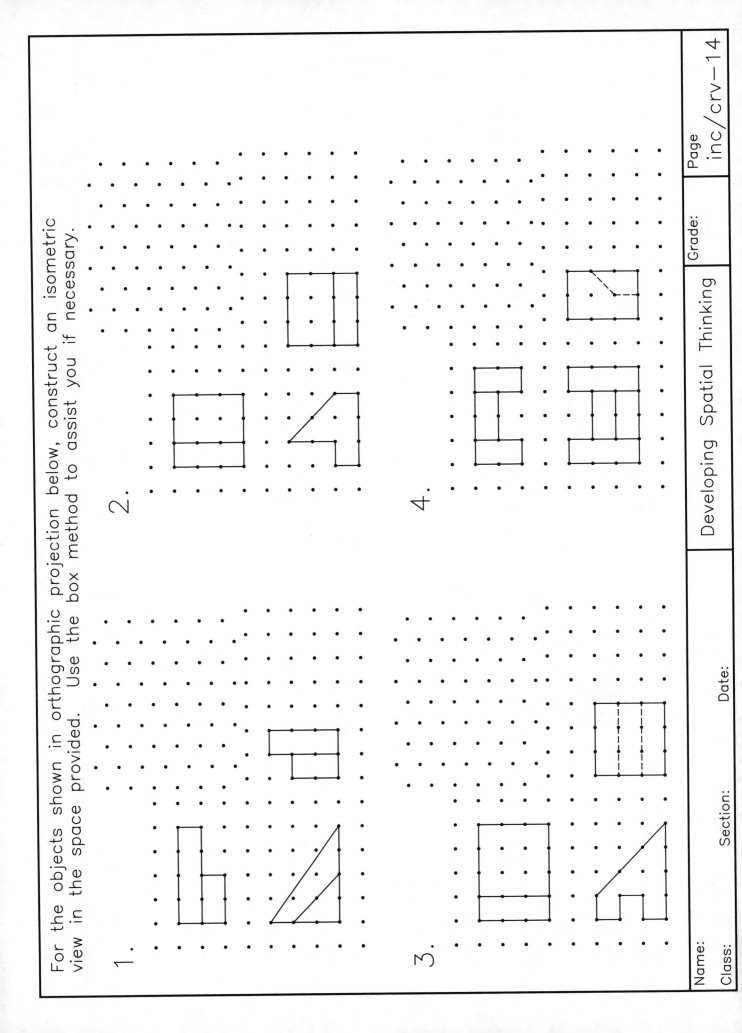

Developing Spatial Thinking

Name:

Class:

Section:

Date:

Grade:

For the objects shown in orthographic projection below, construct an isometric view in the space provided. Use the box method to assist you if necessary.

1.

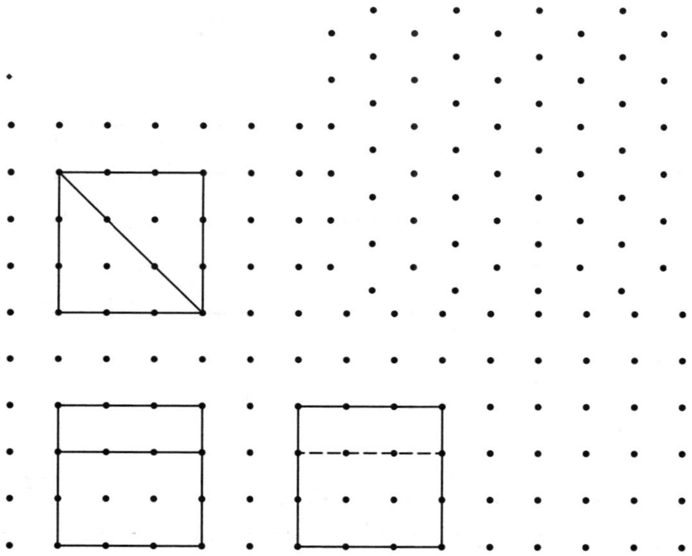

2.

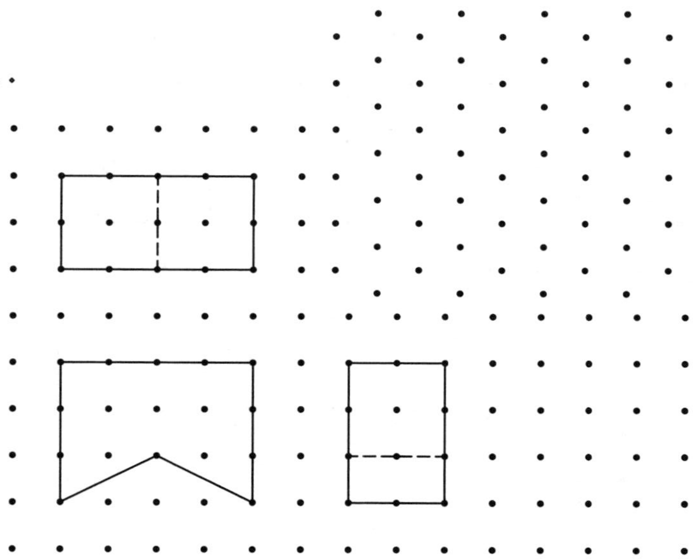

3.

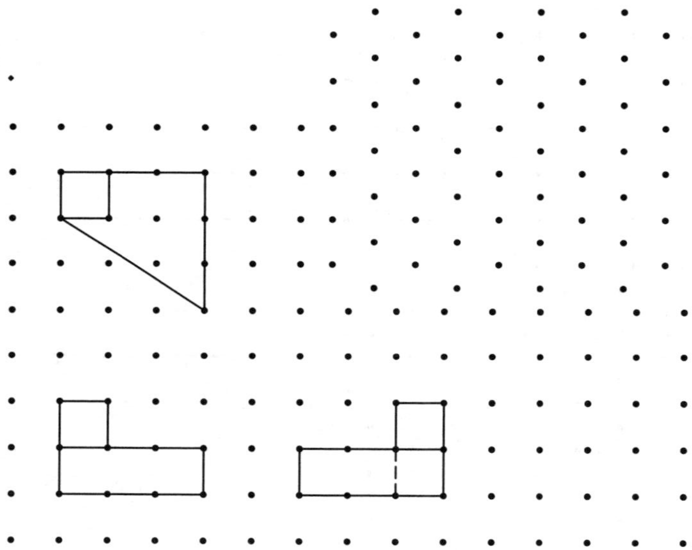

4.

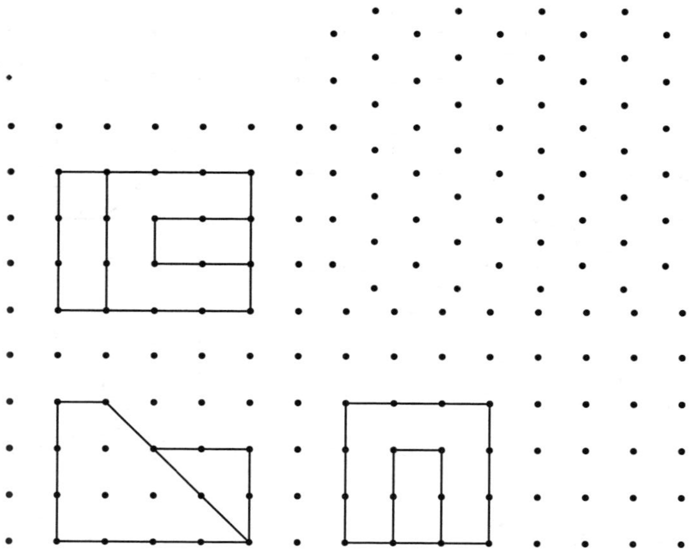

Name:

Class:

Section:

Date:

Grade:

Developing Spatial Thinking

For the objects shown in orthographic projection below, construct an isometric view in the space provided. Use the box method to assist you if necessary.

1.

2.

3.

4.

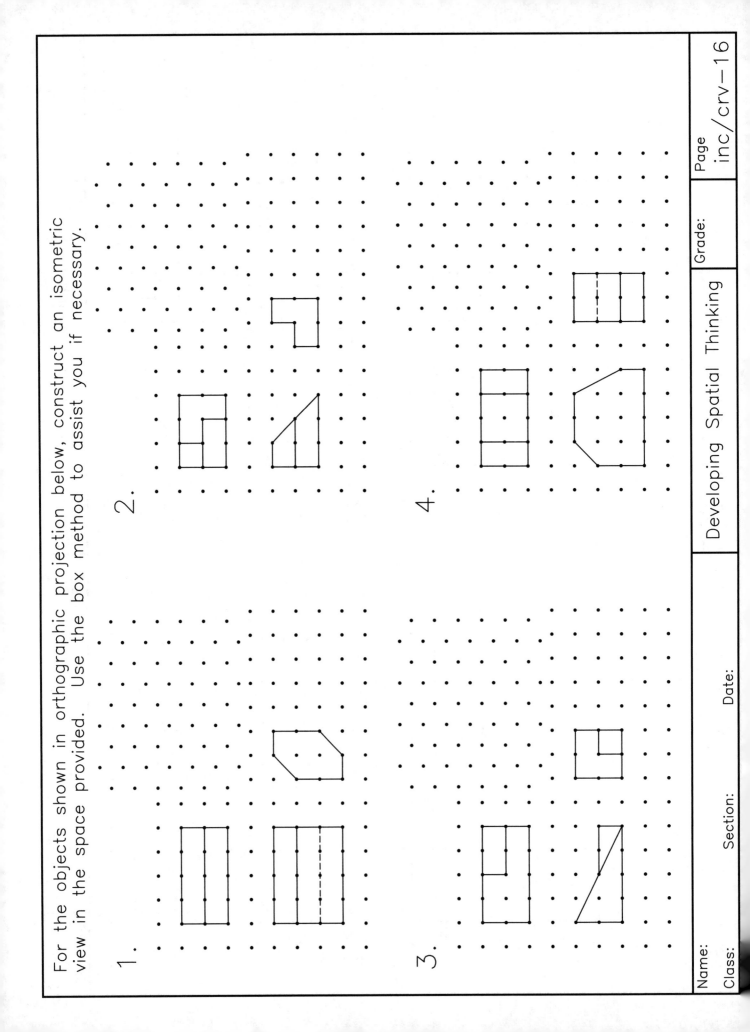

Name:

Class:

Section:

Date:

Developing Spatial Thinking

Page
inc/crv-16

Grade:

Flat Patterns

Many 3-D objects can be formed by folding up a 2-D flat pattern:

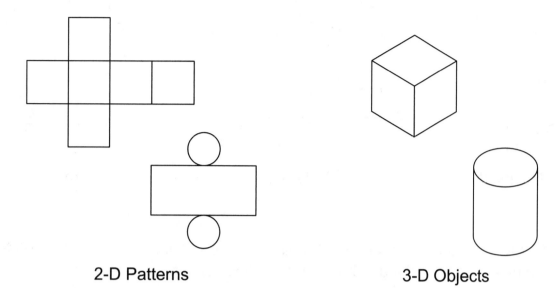

2-D Patterns   3-D Objects

Solid lines on a flat pattern represent fold lines. When constructing a 3-D object from a pattern, fold the pattern along the fold lines.

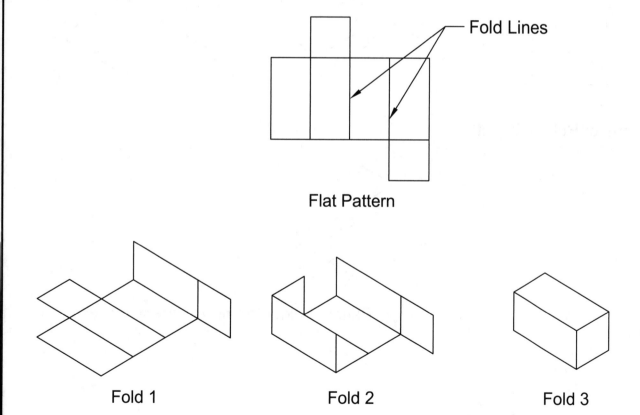

Fold 1   Fold 2   Fold 3

An object can have more than one pattern that can be folded to form it.

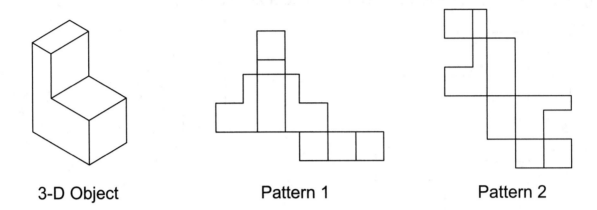

3-D Object   Pattern 1   Pattern 2

Some patterns fold up to form open-ended "hollow" objects. Think of one surface on the pattern as the base and fold along subsequent fold lines until the free edges of the pattern meet.

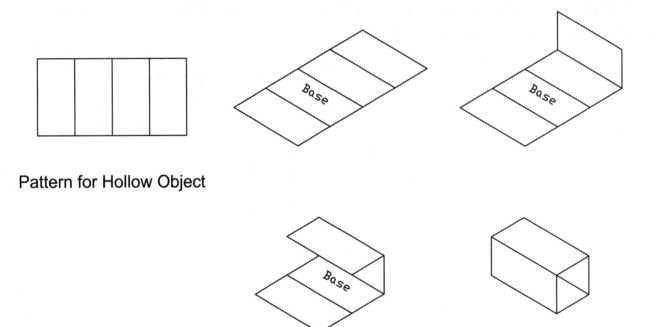

Pattern for Hollow Object

Pattern Folder to Form Hollow Object

Some patterns fold up to form objects with closed ends. Imagine folding up the sides as before, and then fold the ends into place to "close" the object.

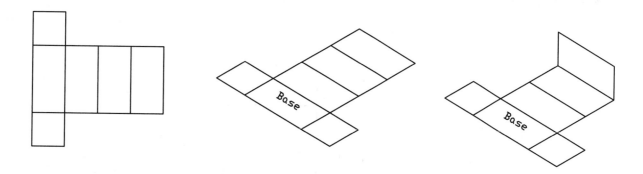

Pattern for Object with Closed Ends

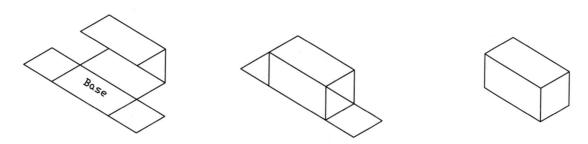

Pattern Folded to Form Closed Object

Sometimes there are makings on a flat pattern. When visualizing folding up a pattern with markings on it, the markings must be oriented on the object the same way they are on the pattern. Also, surfaces that are next to each other on the pattern must be next to each other on the object.

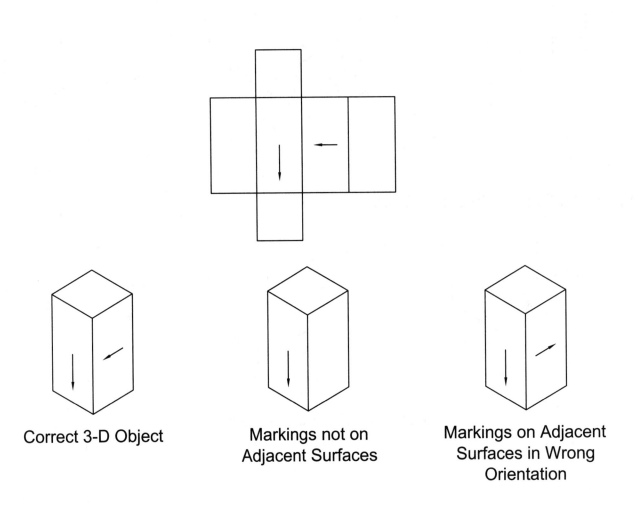

Correct 3-D Object

Markings not on
Adjacent Surfaces

Markings on Adjacent
Surfaces in Wrong
Orientation

The patterns shown below fold up to form a cube with the word "CUBE" spelled around its four sides. Complete each pattern by placing the "B" on it in the correct orientation.

1.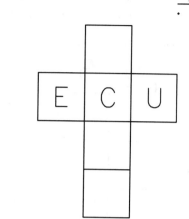

2.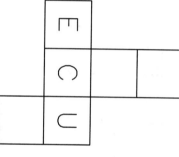

3.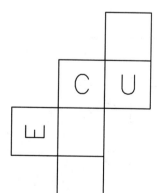

4.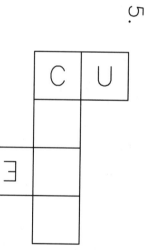

5.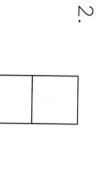

6.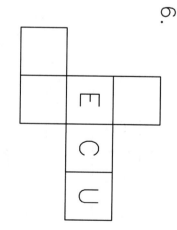

Name:
Class:

Section:

Date:

Developing Spatial Thinking

Grade:

Page
fp—01

The patterns shown below fold up to form a cube with the word "CUBE" spelled around its four sides. Complete each pattern by placing the "B" on it in the correct orientation.

1.

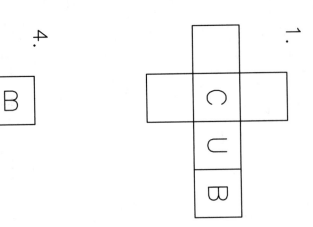

2.

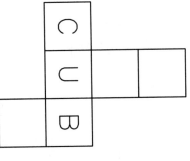

3.

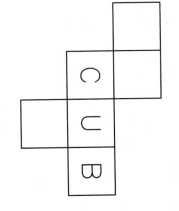

4.

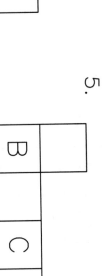

5.

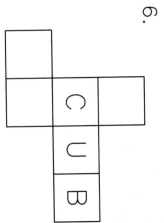

6.

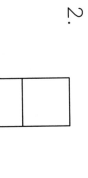

Name: Section: Date:

Class:

Developing Spatial Thinking   Grade: Page
                                                     fp–02

The patterns shown below fold up to form a cube with the word "CUBE" spelled around its four sides. Complete each pattern by placing the "B" on it in the correct orientation.

1.

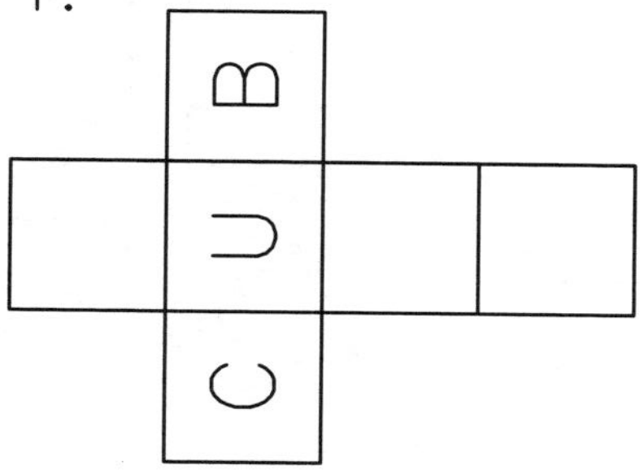

2.

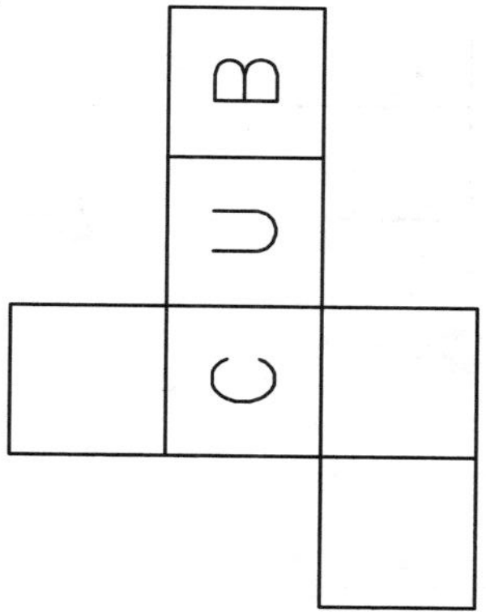

3.

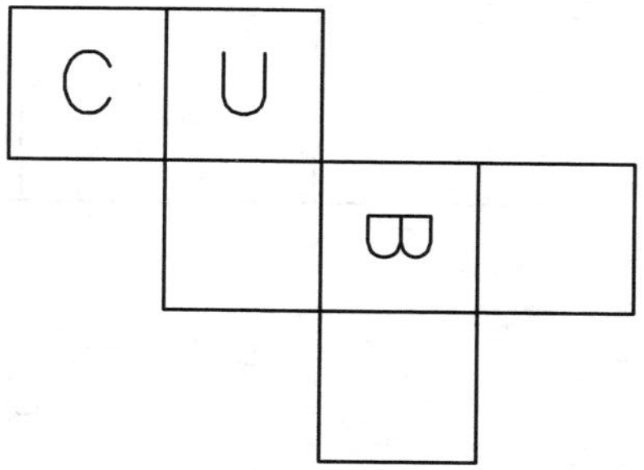

4.

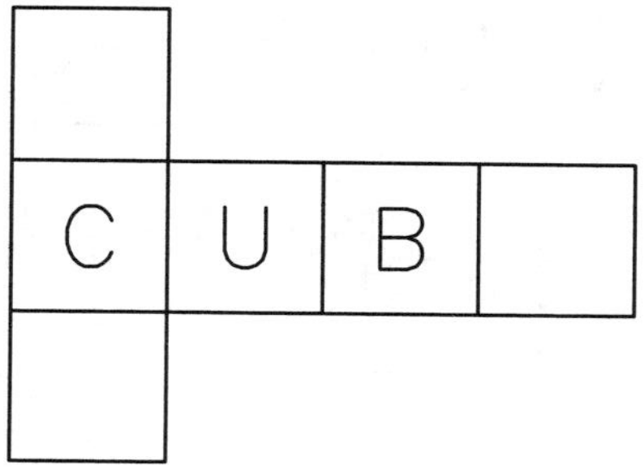

5.

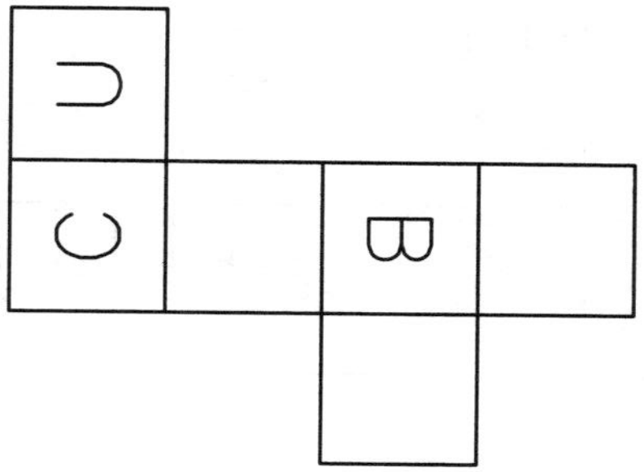

6.

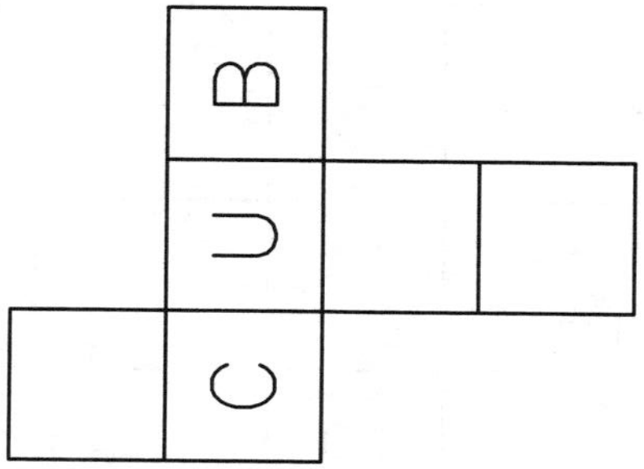

Given the object shown in orthographic projection below, select the letter of the correct flat pattern that could be folded to form it.

1.

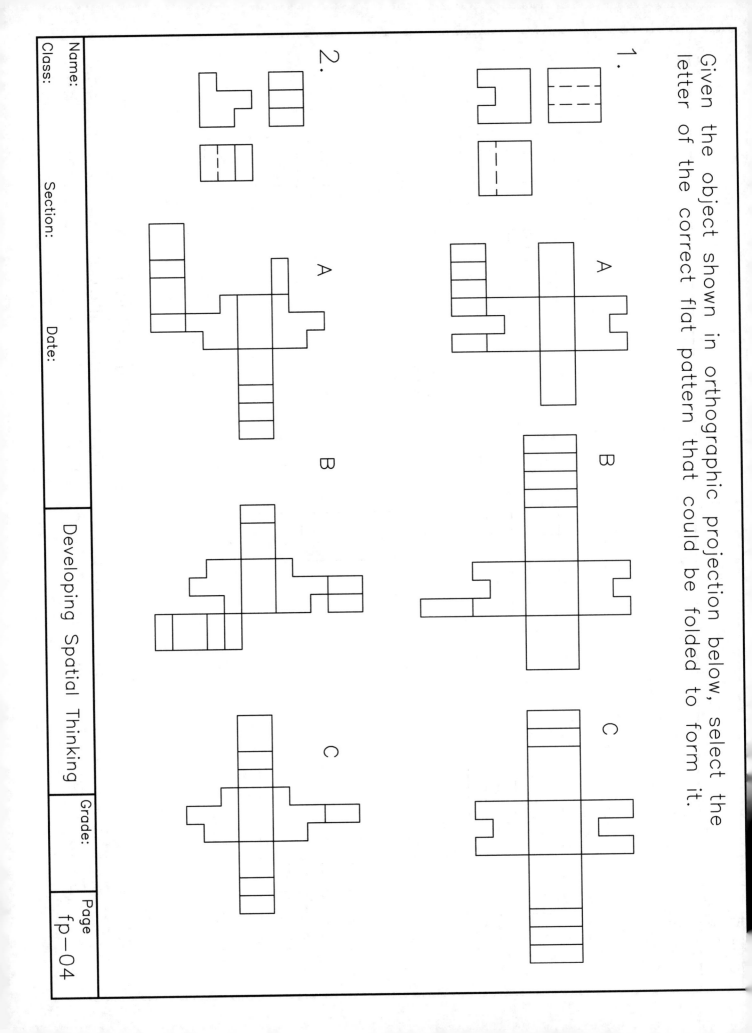

A

B

C

2.

A

B

C

Name: Section: Date:

Class:

Developing Spatial Thinking

Grade:

Page
fp—04

Given the object shown in orthographic projection below, select the letter of the correct flat pattern that could be folded to form it.

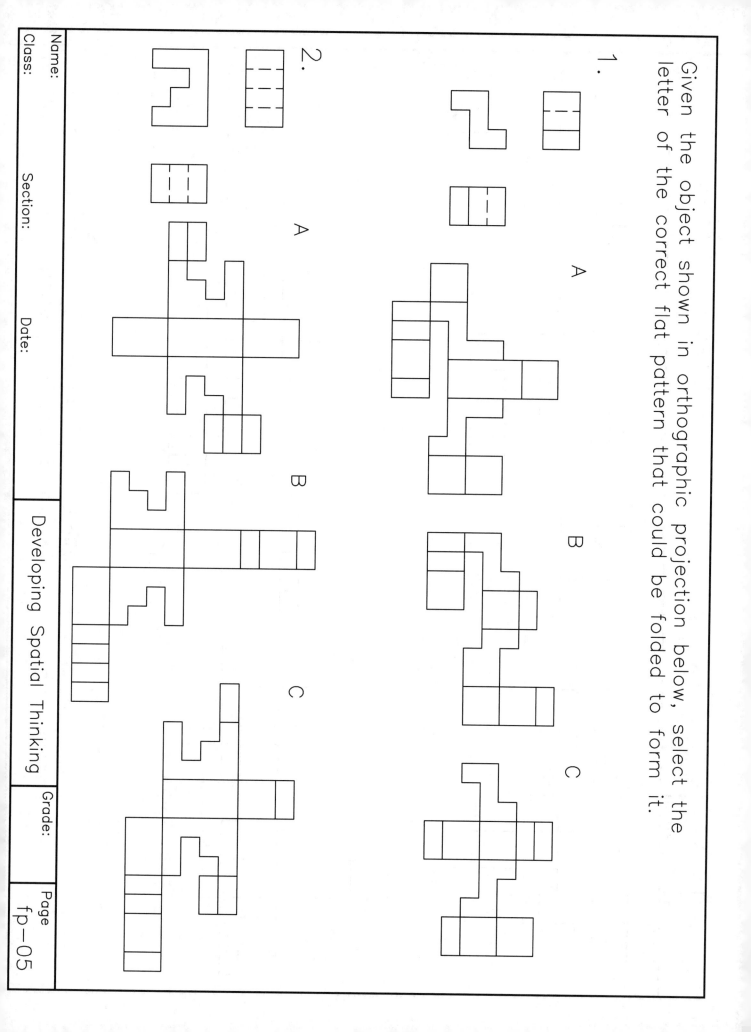

1.

A

B

C

2.

A

B

C

Name:

Class:

Section:

Date:

Developing Spatial Thinking

Grade:

Page
fp—05

Given the object shown in orthographic projection below, select the
letter of the correct flat pattern that could be folded to form it.

1.

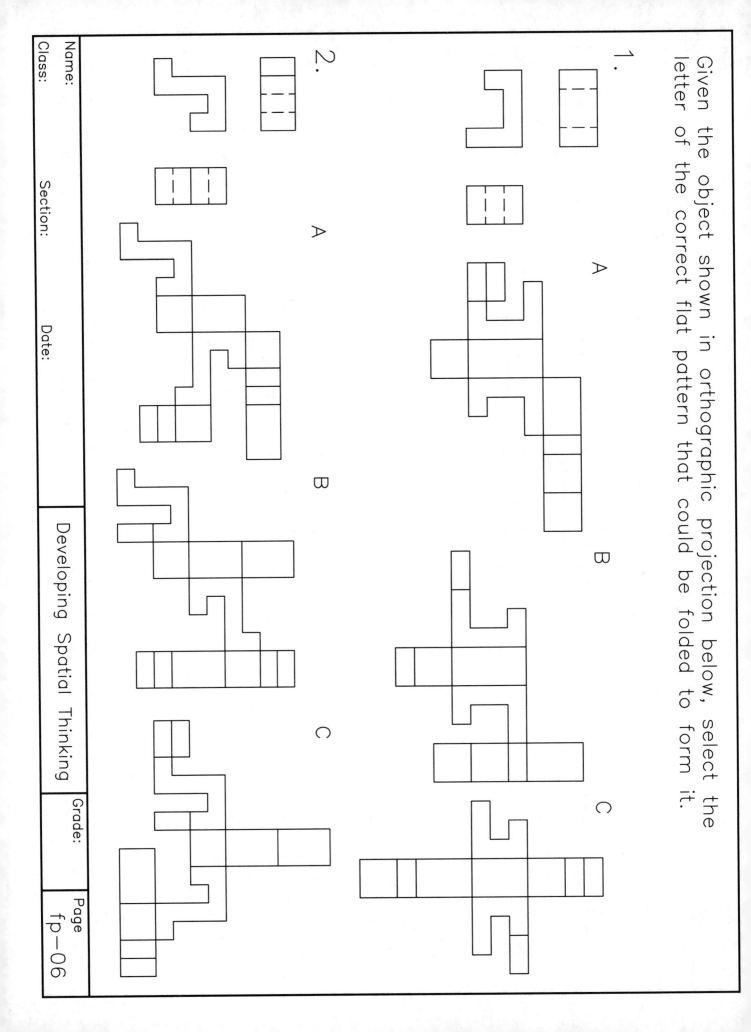

A

B

C

2.

A

B

C

Name:
Class:

Section:

Date:

Developing Spatial Thinking

Grade:

Page
fp—06

Given the object shown in orthographic projection below, select the letter of the correct flat pattern that could be folded to form it.

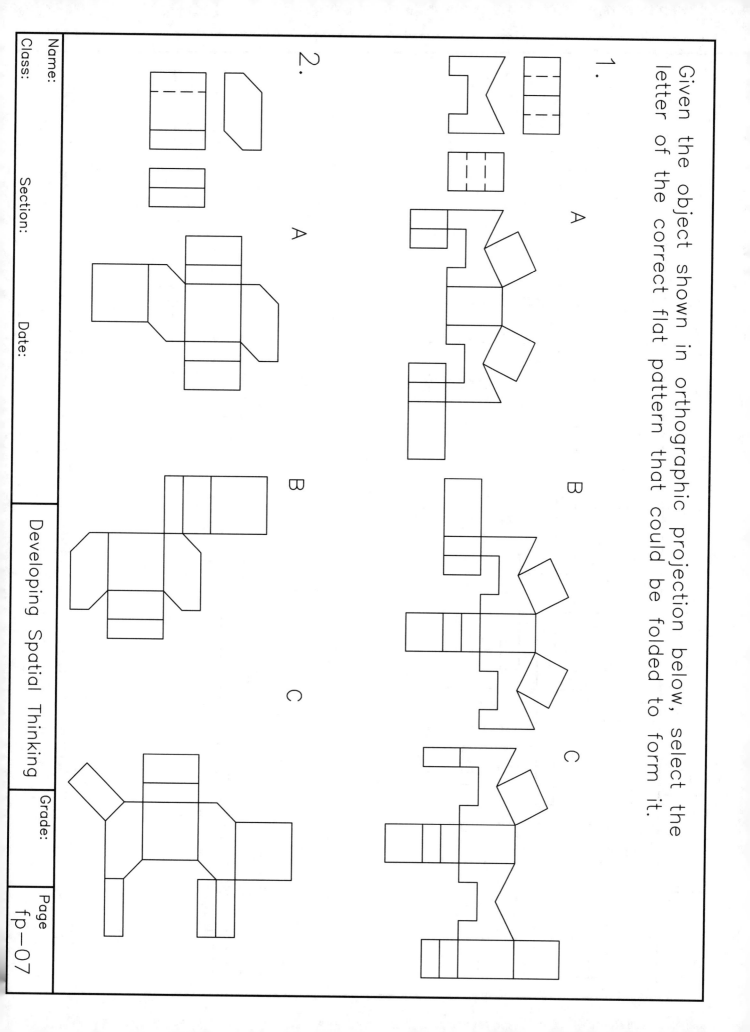

1.

A   B   C

2.

A   B   C

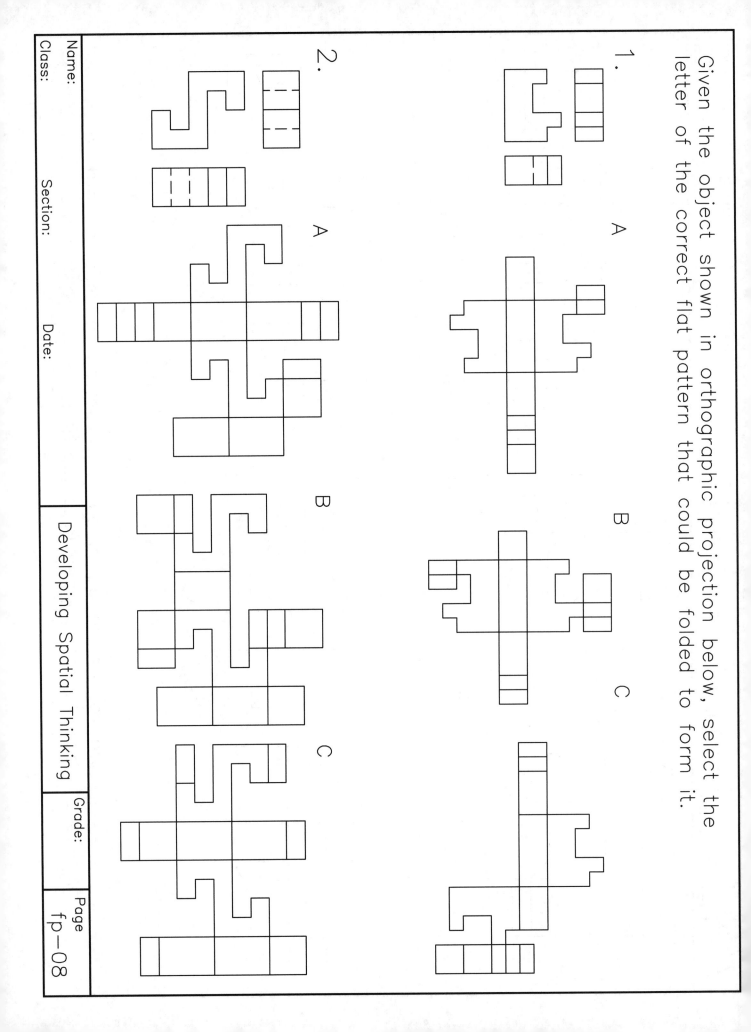

Given the object shown in orthographic projection below, select the letter of the correct flat pattern that could be folded to form it.

1.

A

B

C

2.

A

B

C

Name:

Class:

Section:

Date:

Developing Spatial Thinking

Grade:

Page
fp-08

Select the object that results from folding
up the pattern shown on the left below.

1.

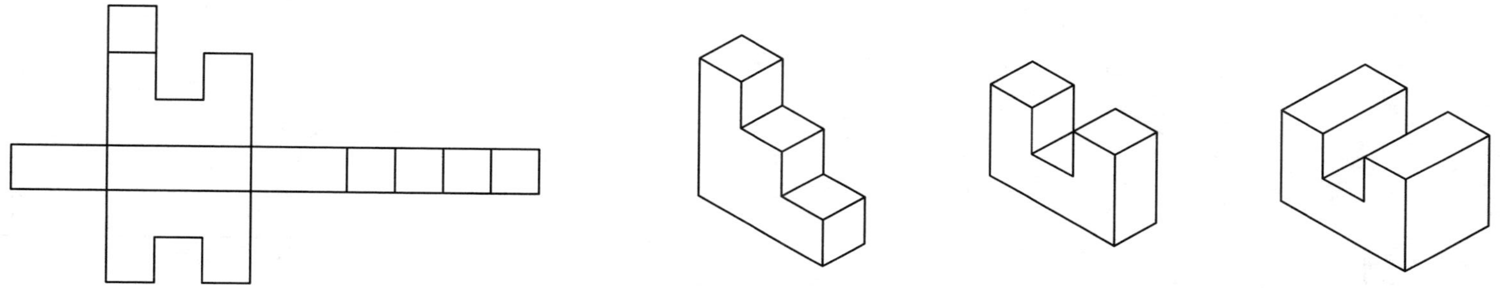

A

B

C

2.

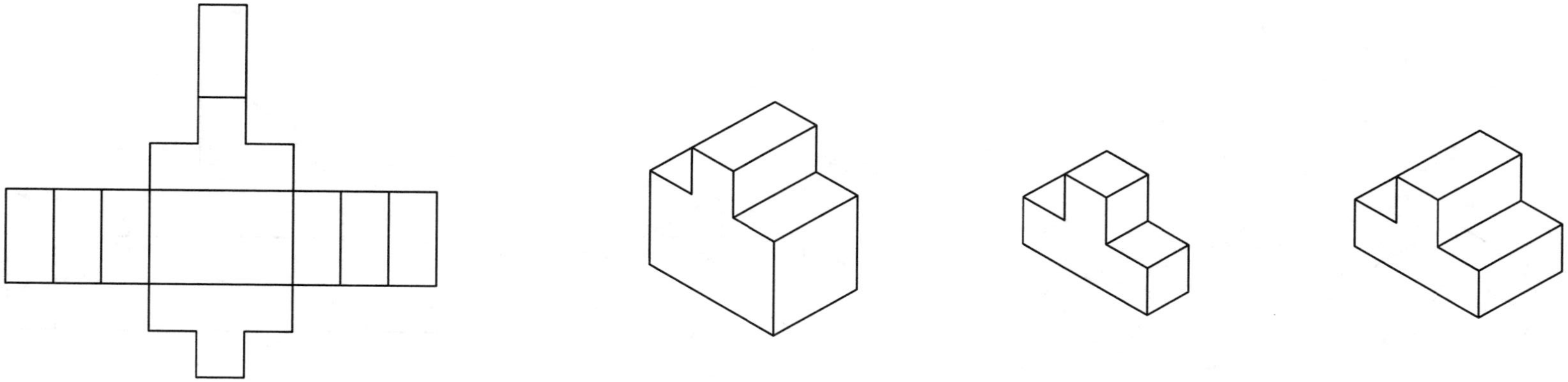

A

B

C

Name: Section: Date:

Class:

Developing Spatial Thinking

Grade:

Page
fp—09

Select the object that results from folding
up the pattern shown on the left below.

1.

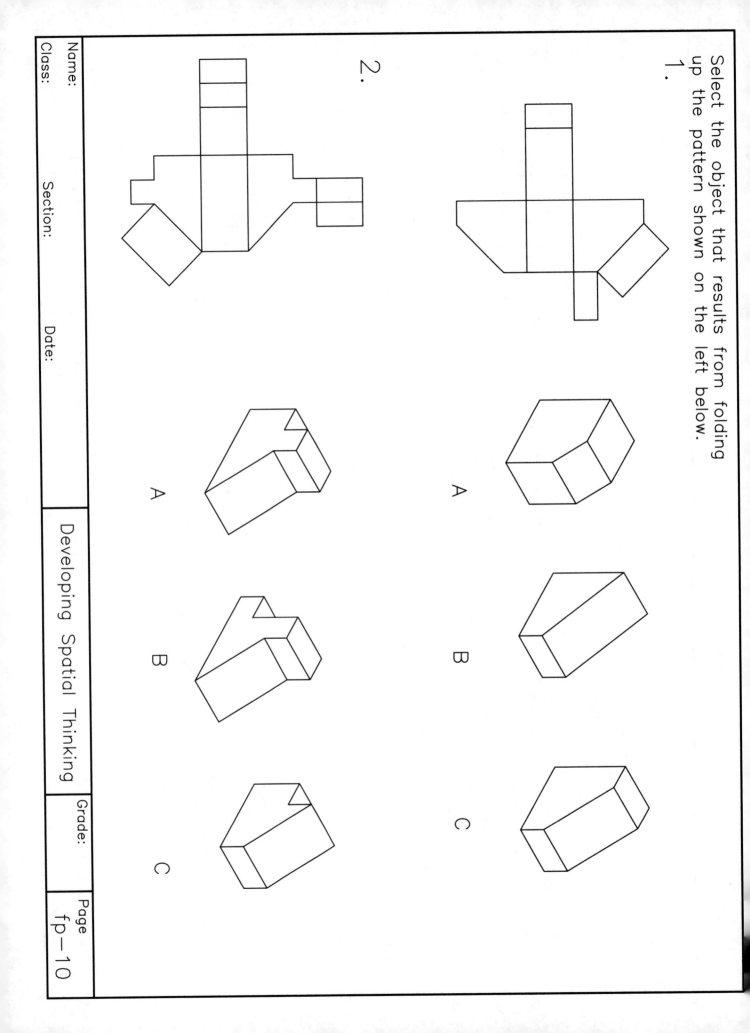

A

B

C

2.

A

B

C

Select the object that results from folding
up the pattern shown on the left below.

1.

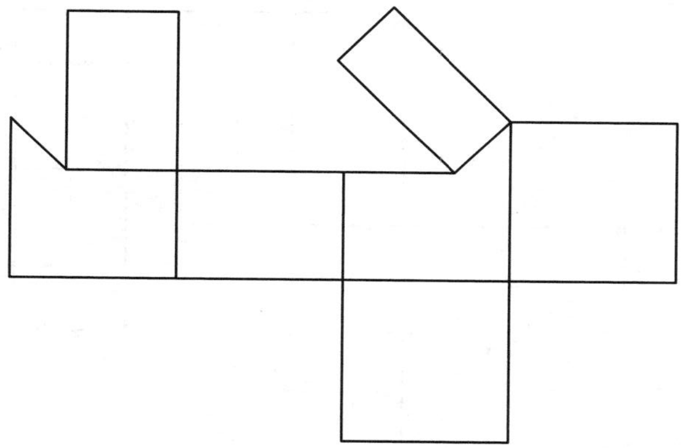

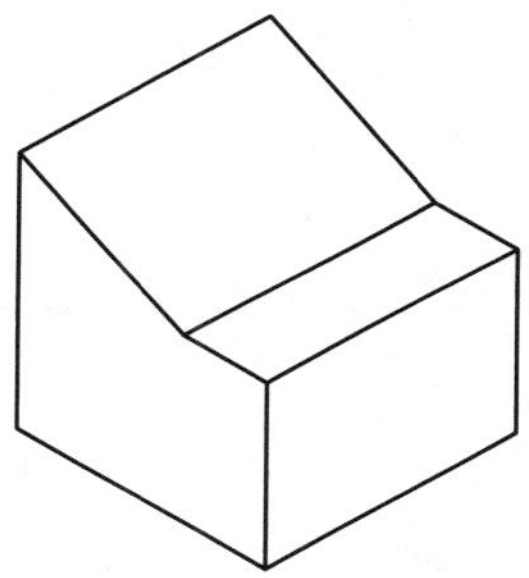

A

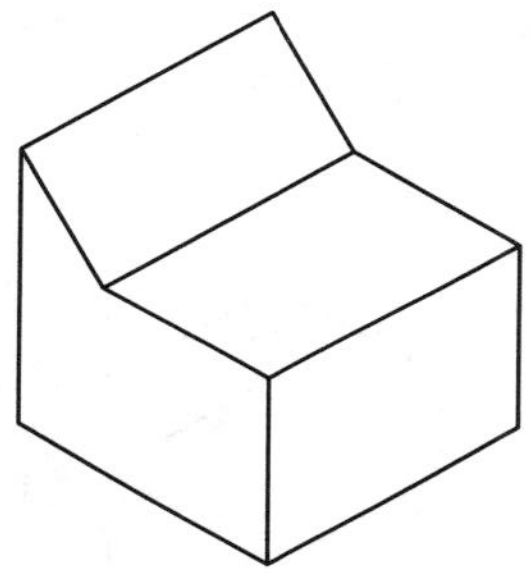

B

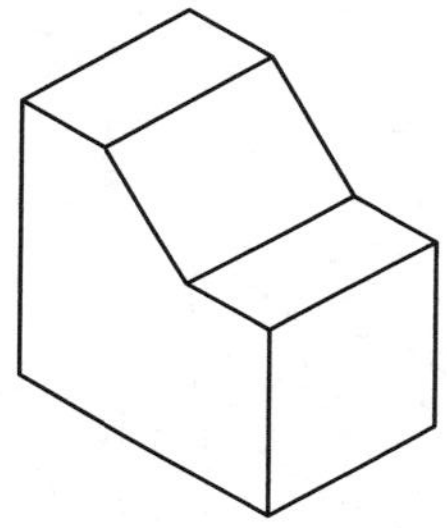

C

2.

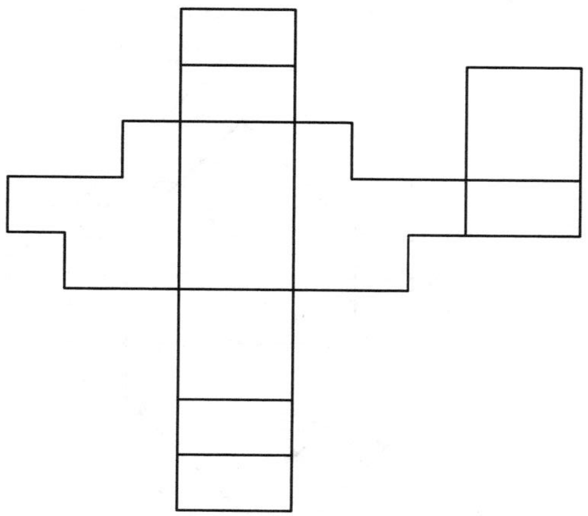

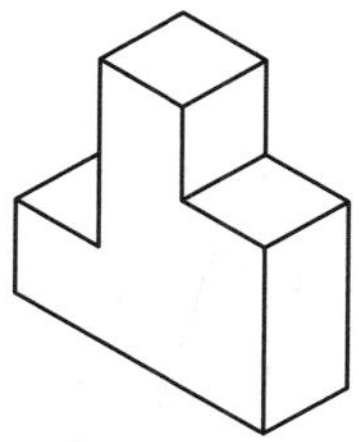

A

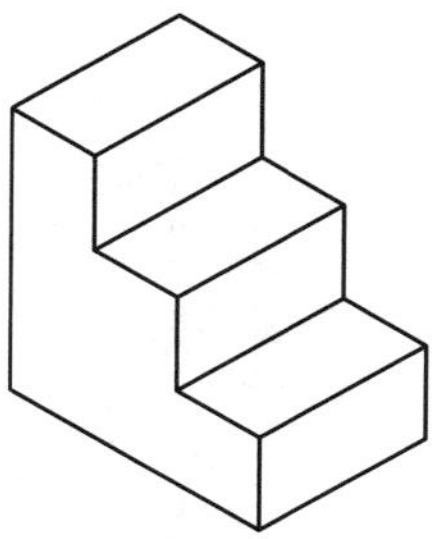

B

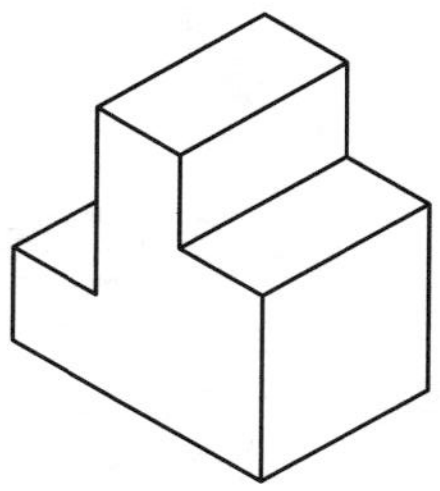

C

Name:

Class:

Section:

Date:

Developing Spatial Thinking

Grade:

Page
fp—11

Select the object that results from folding
up the pattern shown on the left below.

1.

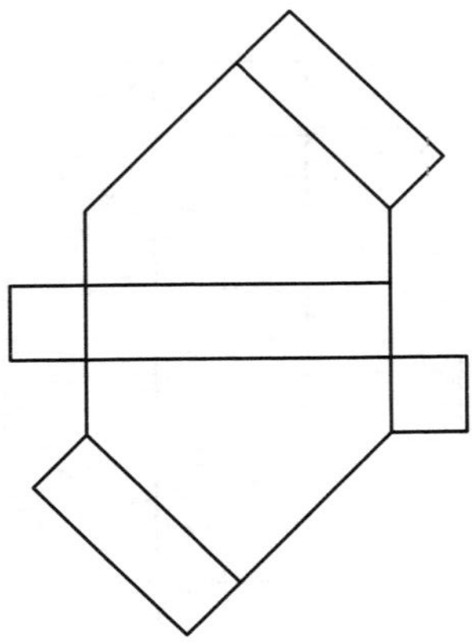

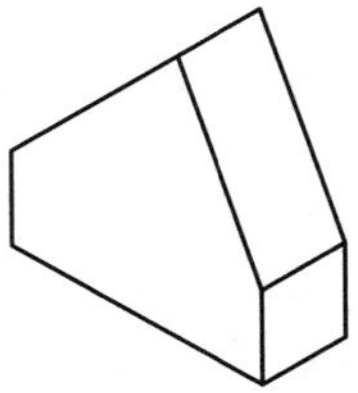

A

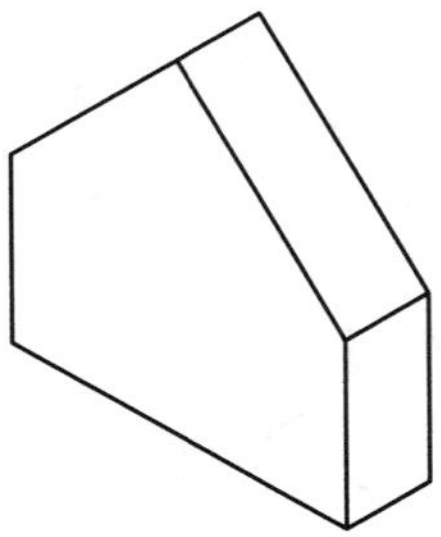

B

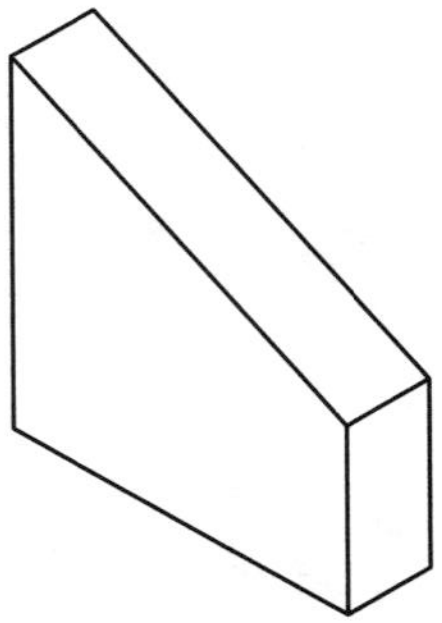

C

2.

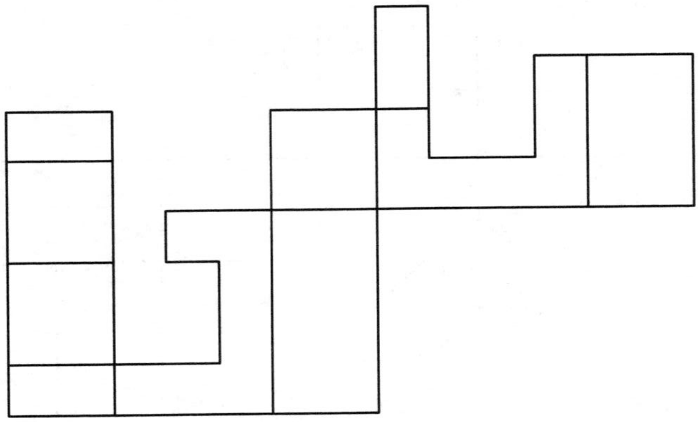

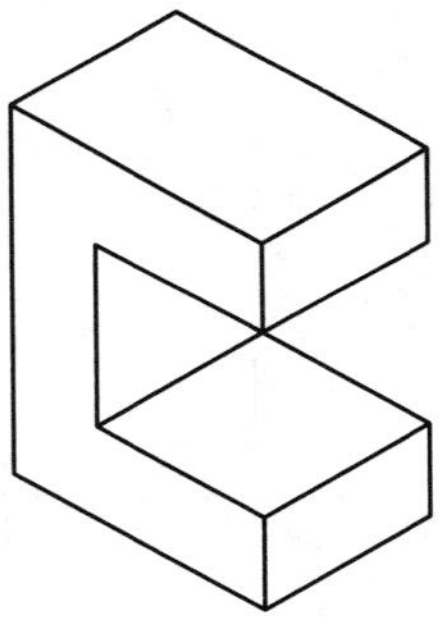

A

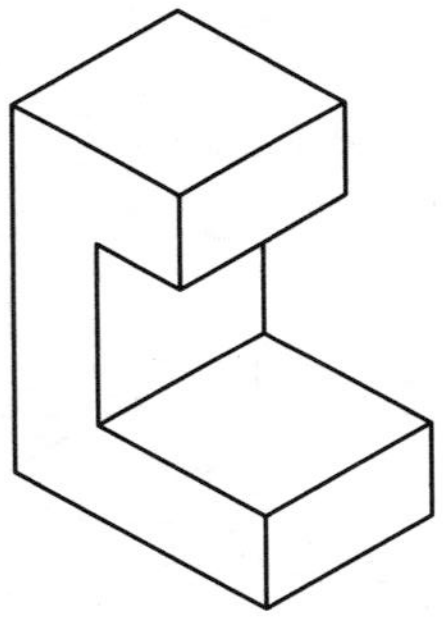

B

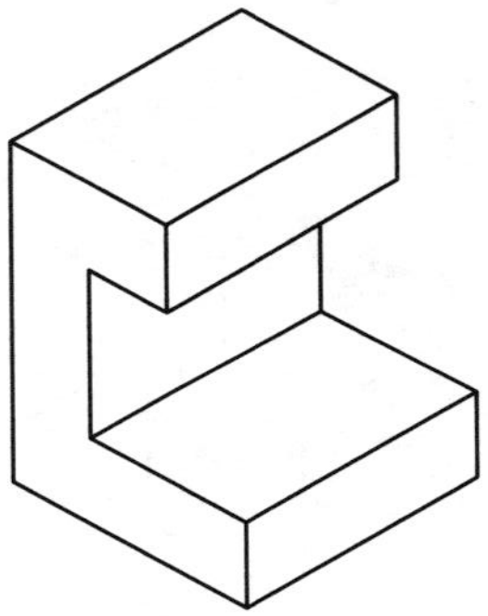

C

Name:

Class:

Section:

Date:

Developing Spatial Thinking

Grade:

Page
fp—12

Select the object that results from folding
up the pattern shown on the left below.

1.

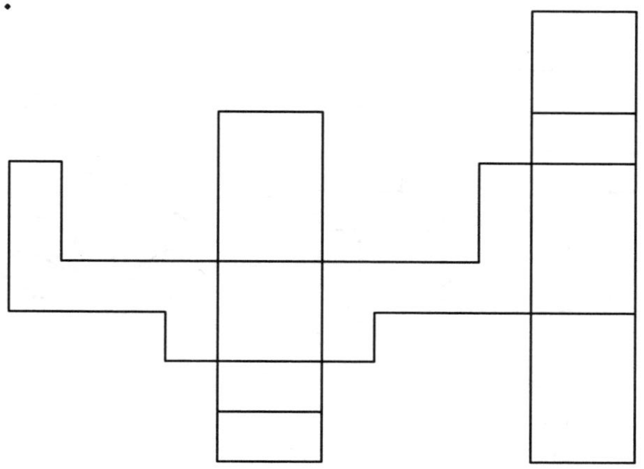

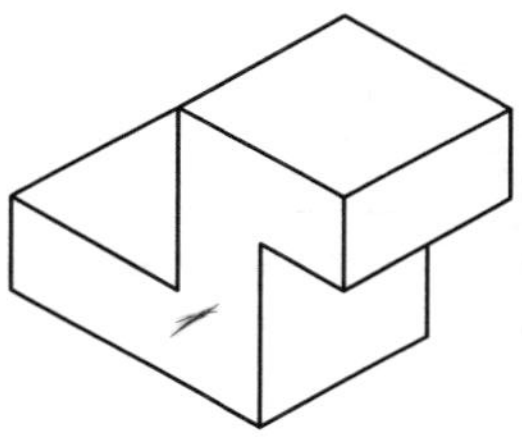

A

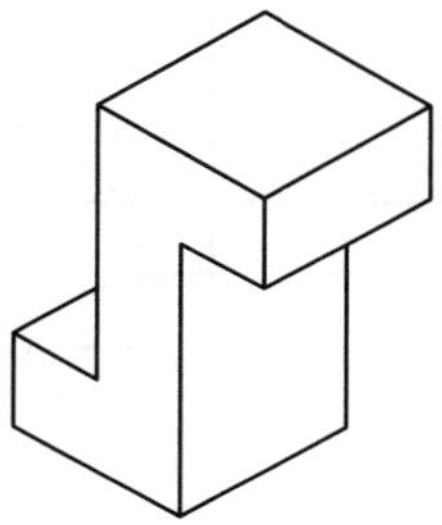

B

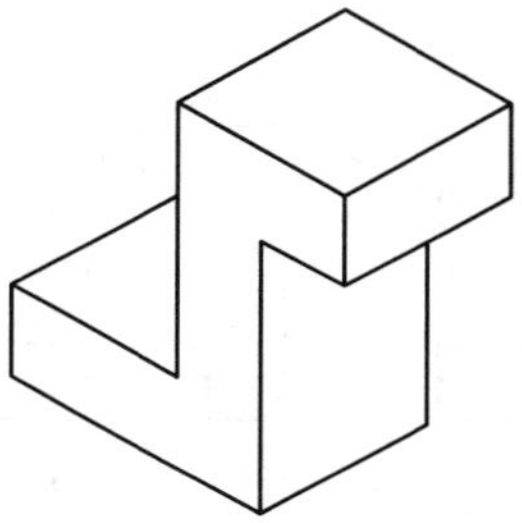

C

2.

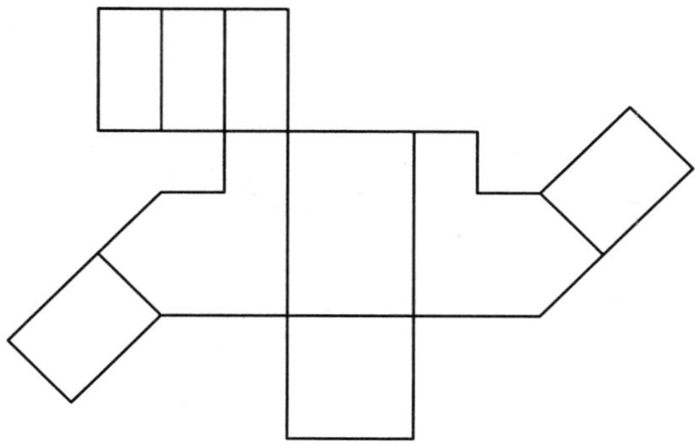

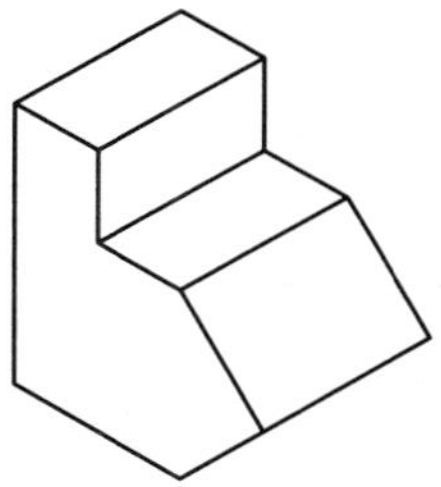

A

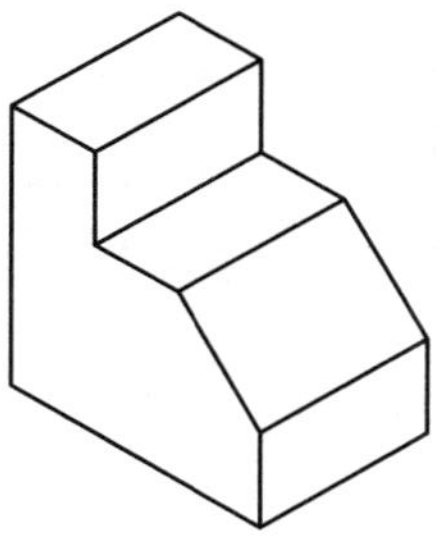

B

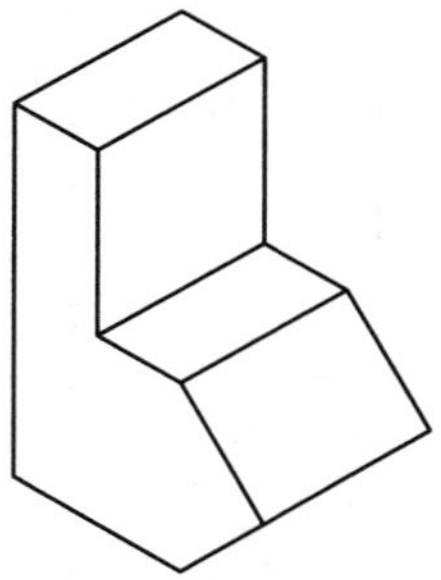

C

For the objects shown in isometric below, select the pattern from the
choices given that could be folded up to obtain the object.

1.

A

B

C

2.

A

B

C

Name:
Class:

Section:

Date:

Developing Spatial Thinking

Grade:

Page
fp−14

For the objects shown in isometric below, select the pattern from the choices given that could be folded up to obtain the object.

1.

A

B

C

2.

A

B

C

Name:

Class:

Section:

Date:

Developing Spatial Thinking

Grade:

Page
fp—15

For the objects shown in isometric below, select the pattern from the choices given that could be folded up to obtain the object.

1.

A

B

C

2.

A

B

C

Name:
Class:

Section:

Date:

Developing Spatial Thinking

Grade:

Page
fp—16

For the objects shown in isometric below, select the pattern from the choices given that could be folded up to obtain the object.

1.

A

B

C

2.

A

B

C

Name:

Class:

Section:

Date:

Developing Spatial Thinking

Grade:

Page
fp–17

For the objects shown in isometric below, select the pattern from the choices given that could be folded up to obtain the object.

1.

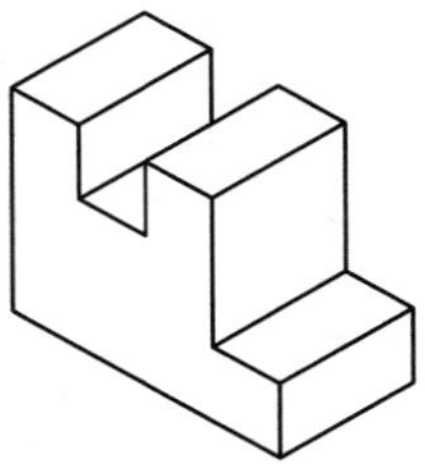

A

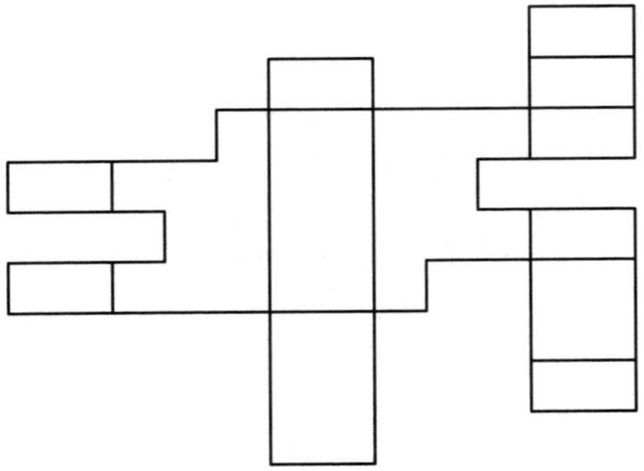

B

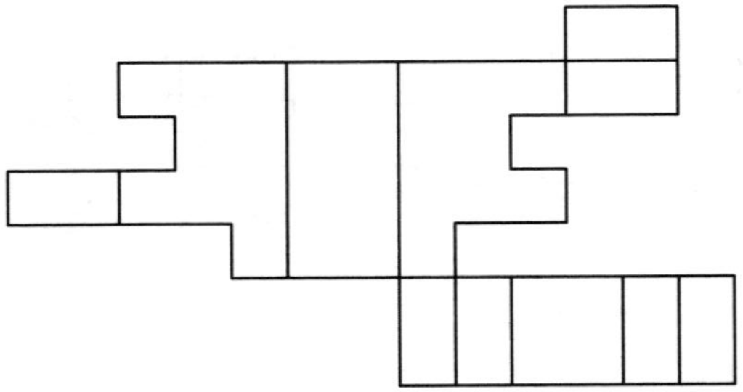

C

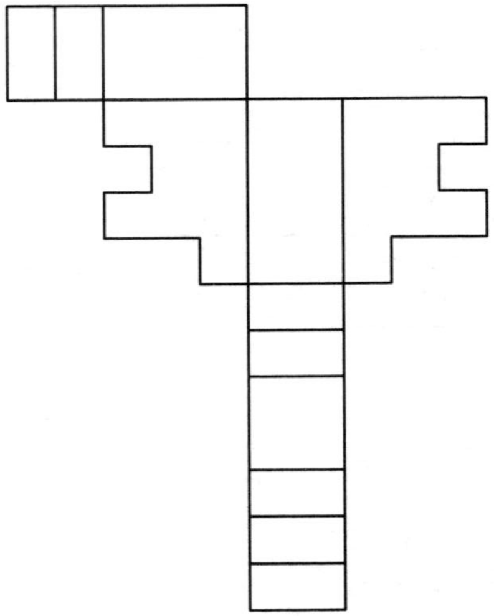

2.

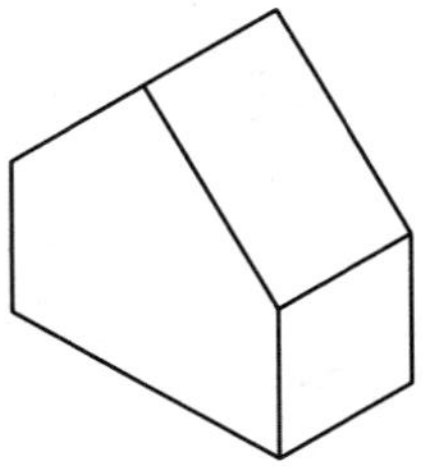

A

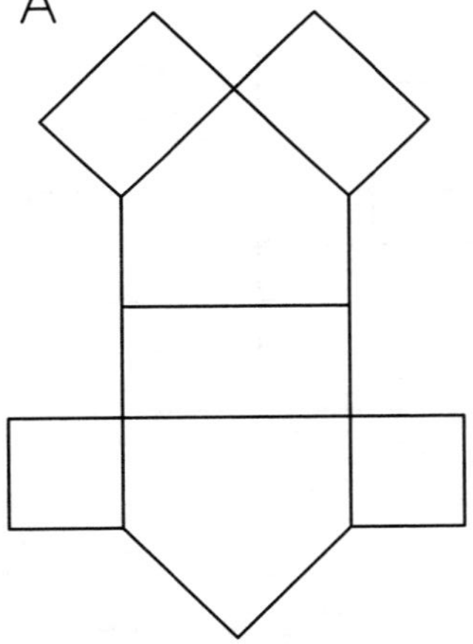

B

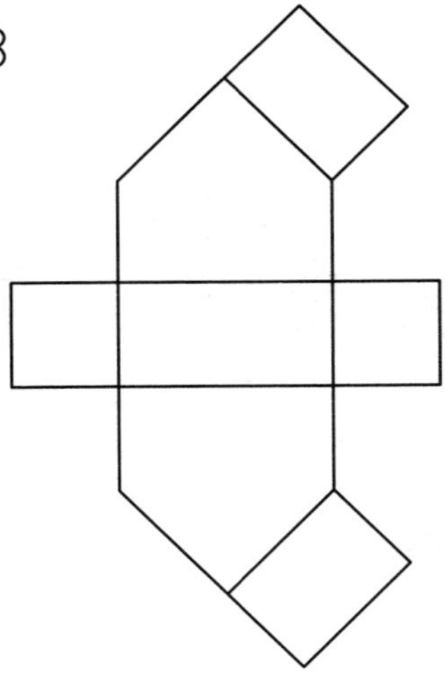

C

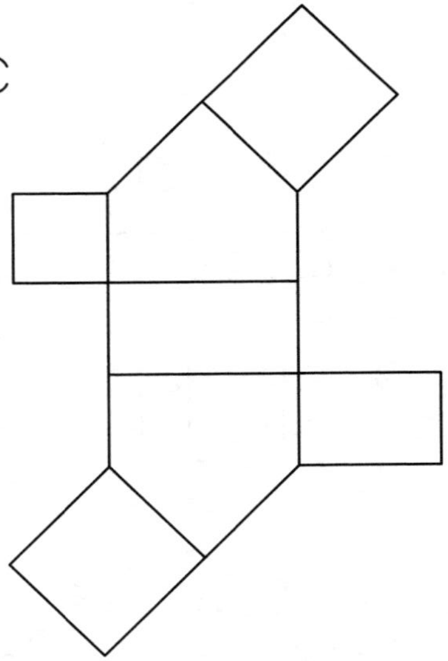

The rectangular prisms shown in isometric below have markings on each of their four long sides. Choose the panel for the flat pattern so that when folded up, the correct object is obtained.

1.

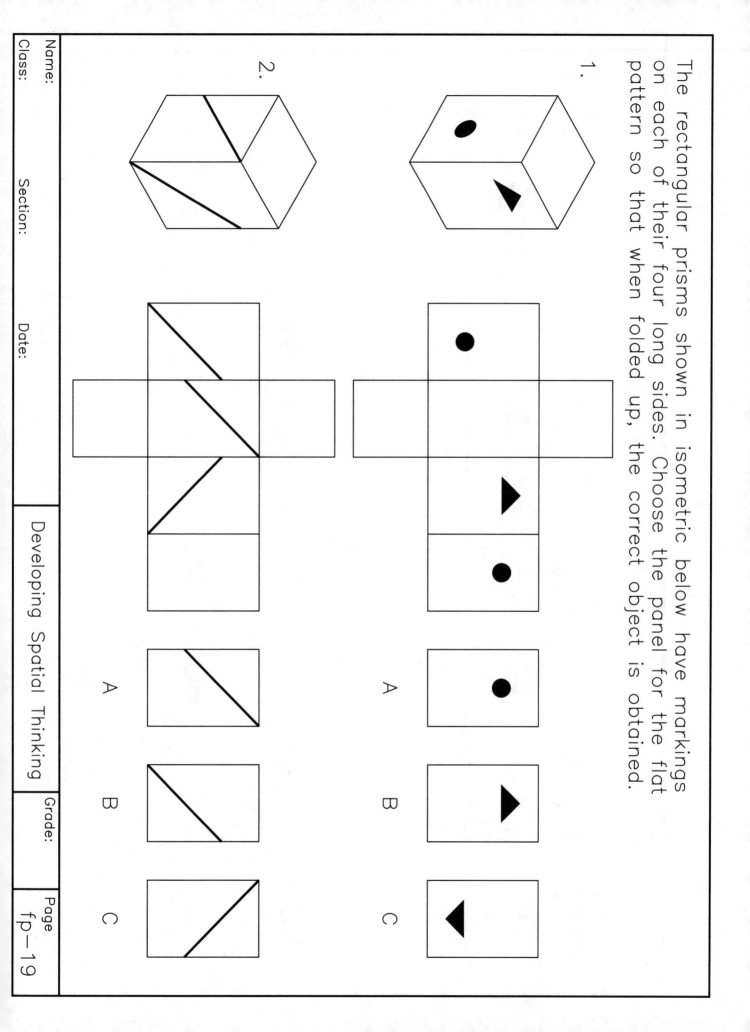

A   B   C

2.

A   B   C

The rectangular prisms shown in isometric below have markings on each of their four long sides. Choose the panel for the flat pattern so that when folded up, the correct object is obtained.

1.

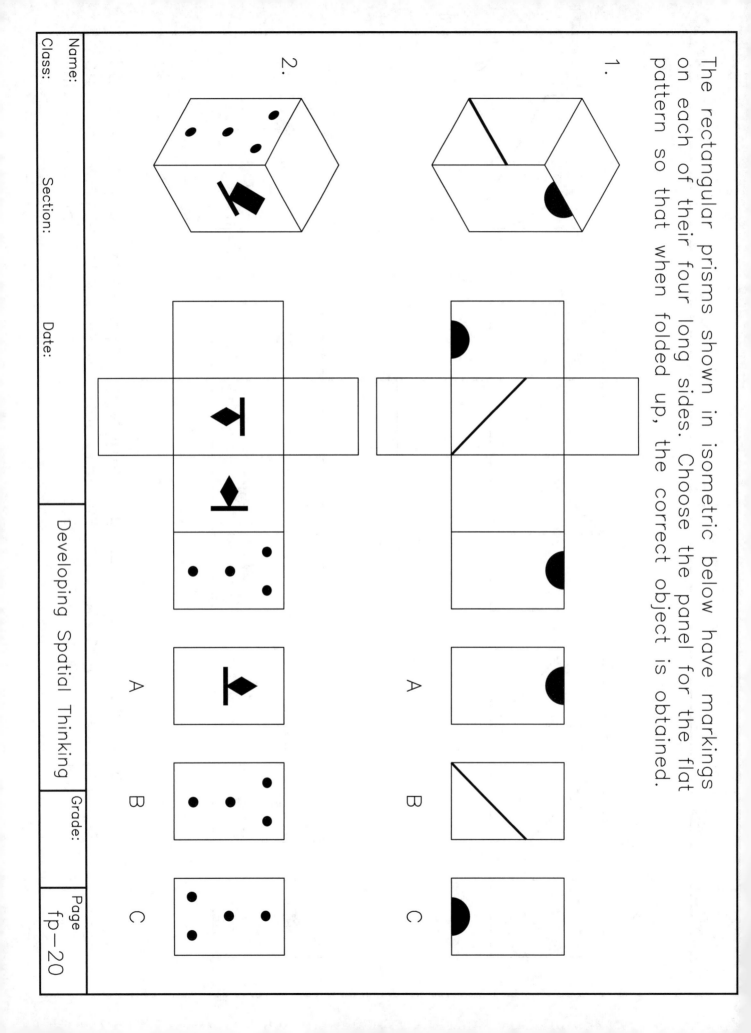

A

B

C

2.

A

B

C

The rectangular prisms shown in isometric below have markings on each of their four long sides. Choose the panel for the flat pattern so that when folded up, the correct object is obtained.

1.

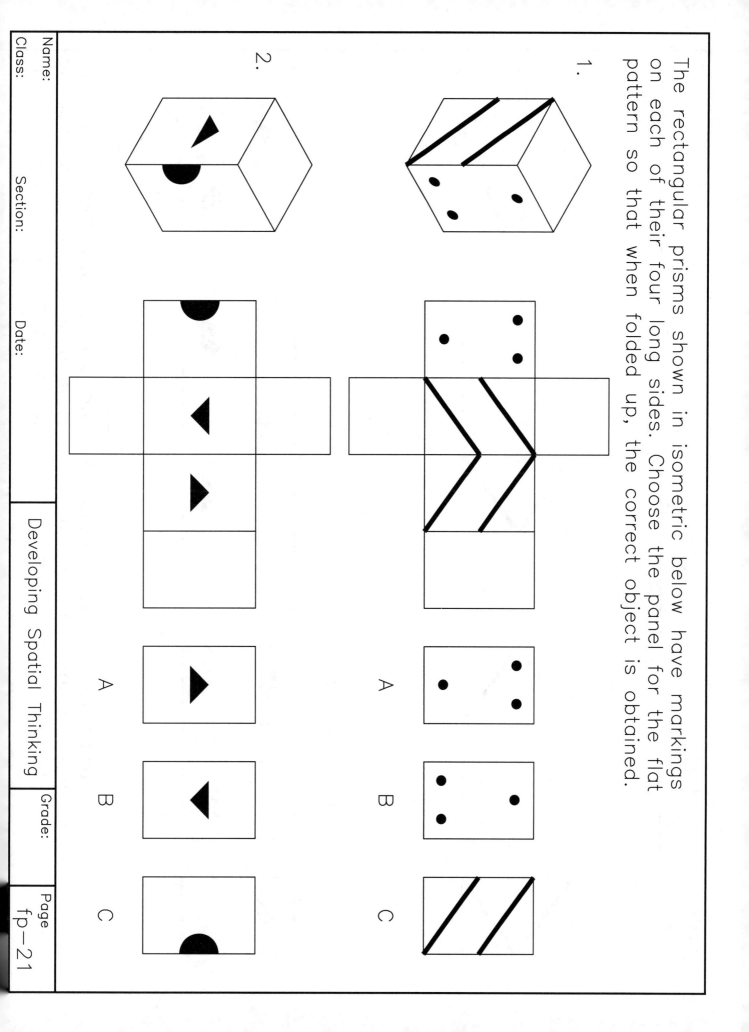

A   B   C

2.

A   B   C

The rectangular prisms shown in isometric below have markings on each of their four long sides. Choose the panel for the flat pattern so that when folded up, the correct object is obtained.

1.

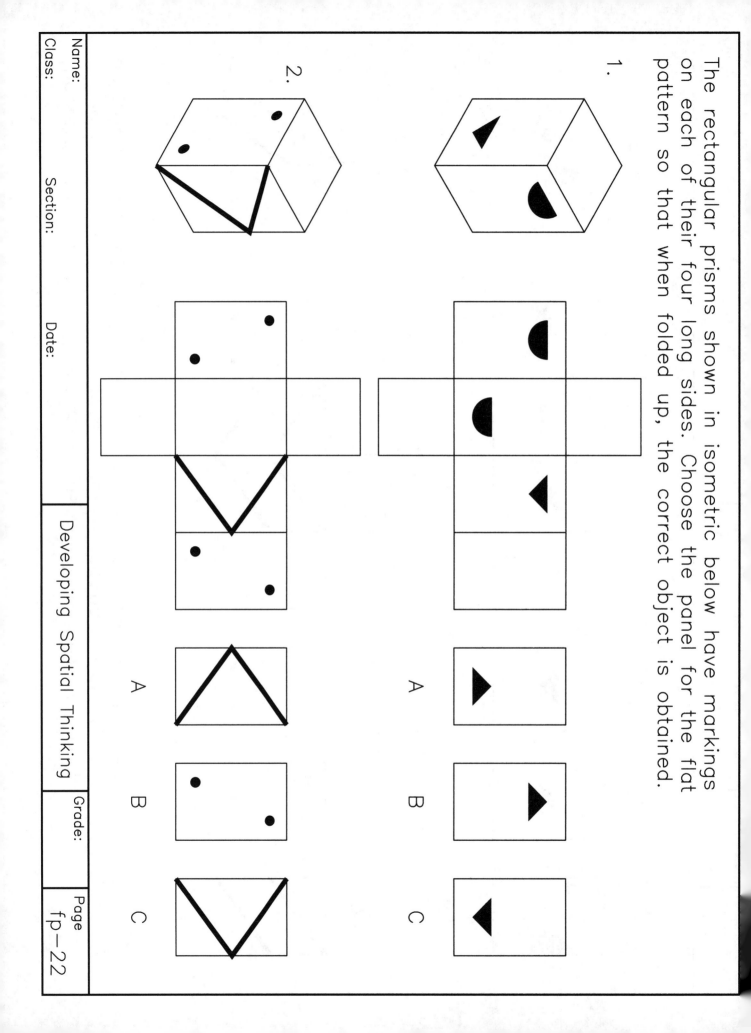

A   B   C

2.

A   B   C

The rectangular prisms shown in isometric below have markings on each of their four long sides. Choose the panel for the flat pattern so that when folded up, the correct object is obtained.

1.

A

B

C

2.

A

B

C

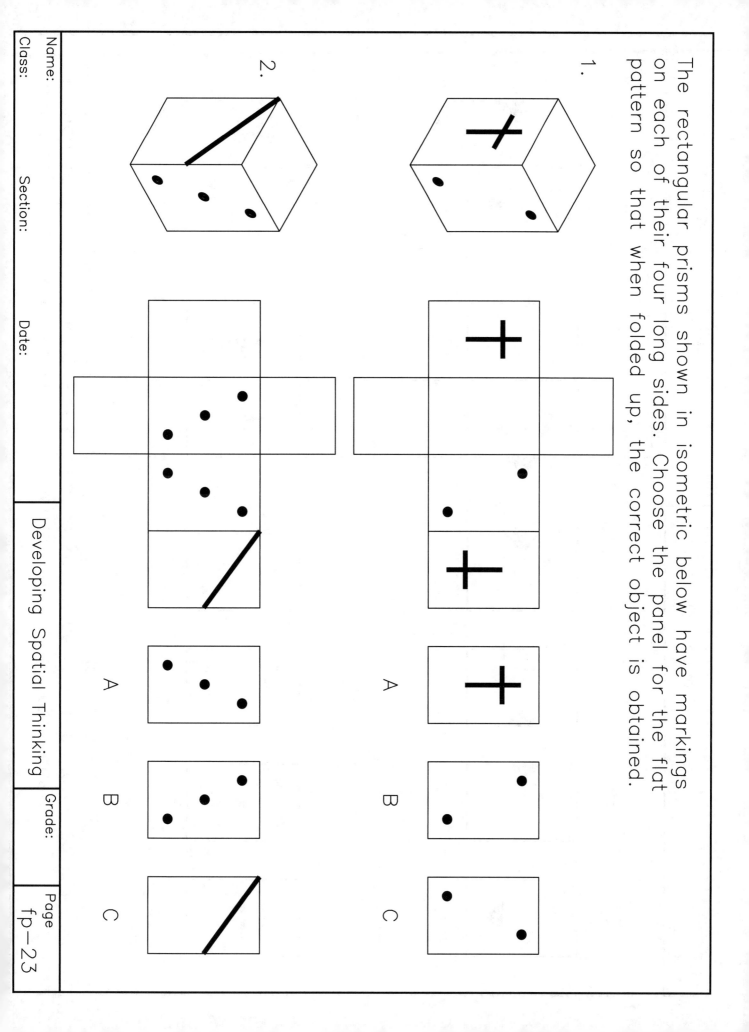

Name:
Class:

Section:

Date:

Developing Spatial Thinking

Grade:

Page
fp—23

The rectangular prisms shown in isometric below have markings on each of their four long sides. Choose the panel for the flat pattern so that when folded up, the correct object is obtained.

1.

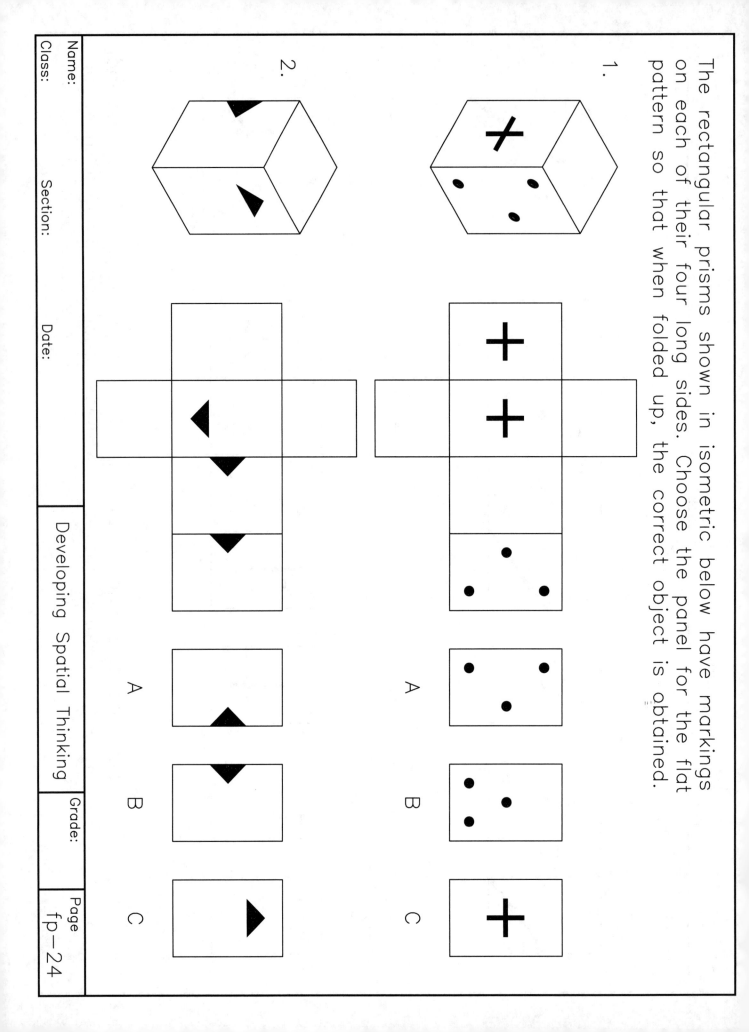

A

B

C

2.

A

B

C

Rotation of Objects About a Single Axis

A rotation of an object is a turning of it about a straight line. The line about which the object rotates is called the *axis of rotation.*

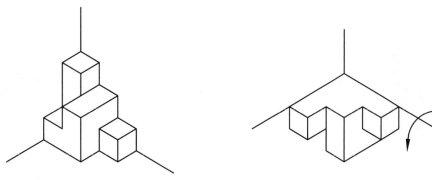

Original Object Position

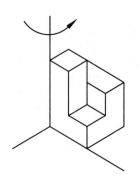

An object can rotate either positivey or negatively about an axis. If you look down the axis of rotation, a *positive* rotation is counterclockwise and a *negative* rotation is clockwise.

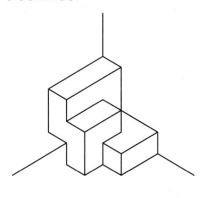

Original Object Position

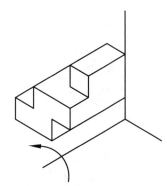

Positive Rotation

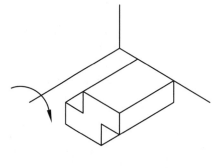

Negative Rotation

The direction of the rotation is determined by the right hand rule. For a positive rotation, if you point the thumb of your right hand along the positive direction of the axis of rotation, your fingers will curl in the direction of the rotation. For a negative rotation, if you point the thumb of your right hand along the negative axis of rotation, your fingers will curl in the direction of the rotation.

Positive Rotation

Negative Rotation

Object rotation can be represented by the following rotation designation coding scheme: A single rotation arrow designation represents a 90 degree rotation about an axis. An arrow in the counterclockwise direction indicates a positive rotation and an arrow in the clockwise direction indicates a negative rotation.

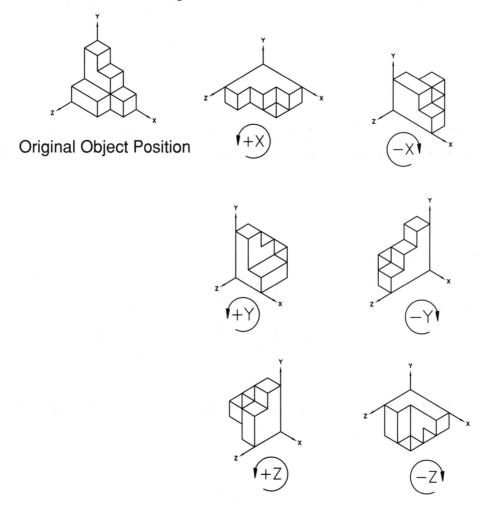

Original Object Position

An object can be rotated about an axis multiple times. For each increment of 90 degrees of rotation, a new arrow is included in the rotation designation coding scheme.

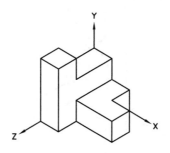

Original Object Position

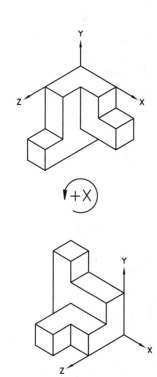

$\left(+X\right)$

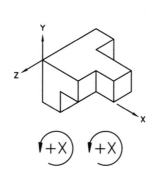

$\left(+X\right)$ $\left(+X\right)$

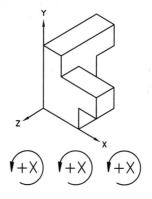

$\left(+X\right)$ $\left(+X\right)$ $\left(+X\right)$

$\left(-Y\right)$

$\left(-Y\right)$ $\left(-Y\right)$

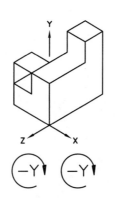

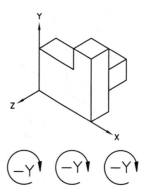

$\left(-Y\right)$ $\left(-Y\right)$ $\left(-Y\right)$

Different combinations of rotations can produce the same result. Thus, one arrow rotation designation sequence can be replaced by another one because the two are equivalent.

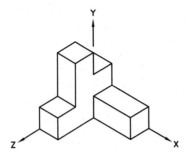

Original Object Position

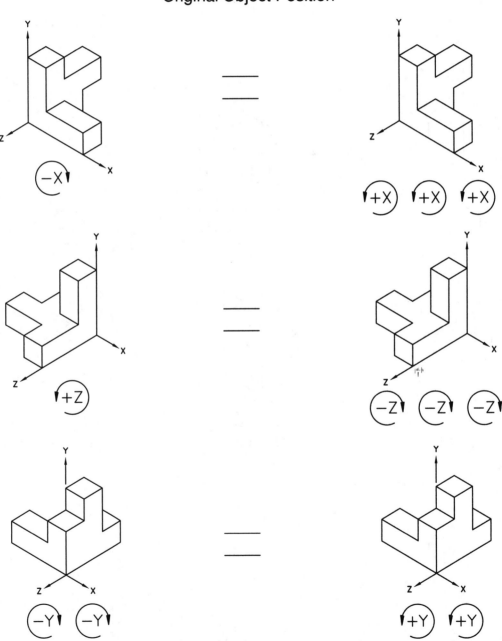

One way to think about rotation direction is to transfer the given rotation to the positive end of the axis in question. For example, for the objects shown below with the specified rotations - if you move the rotations to the end of the axis, the curved arrow shows you the direction of the rotation.

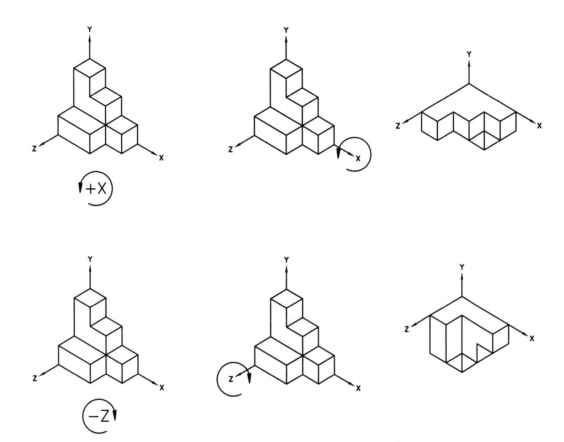

For the objects shown below sketch the object in the space provided after rotating it about the axis by the indicated amount.

1.

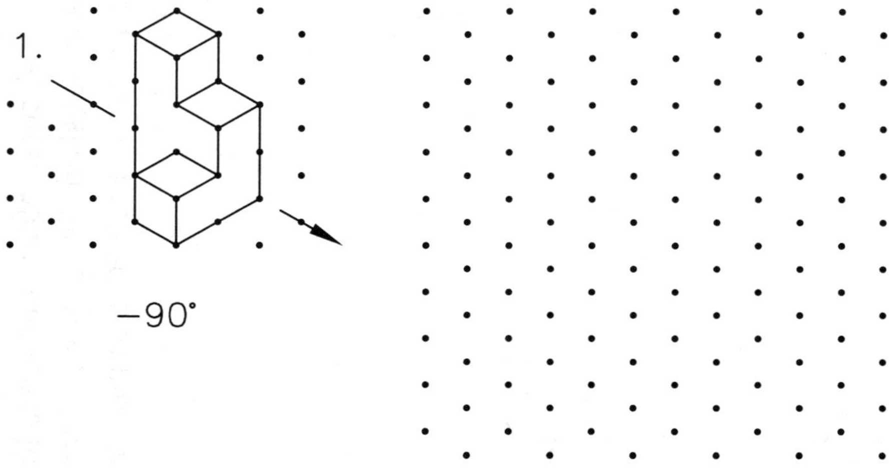

−90°

2.

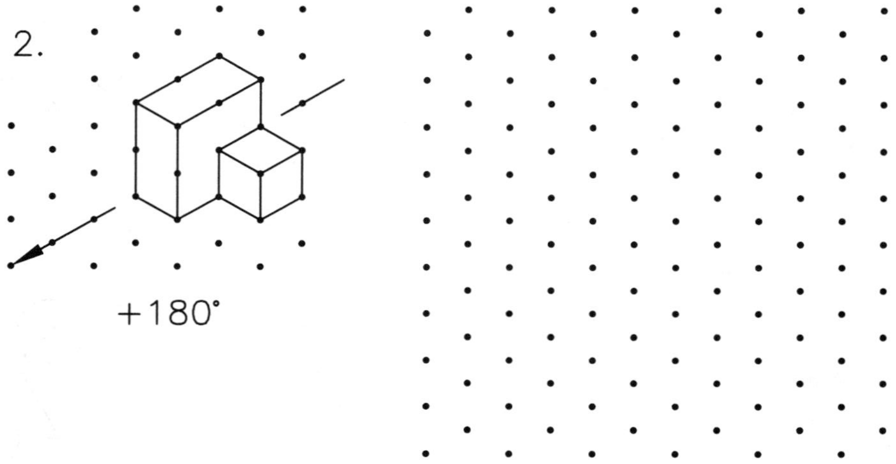

+180°

3.

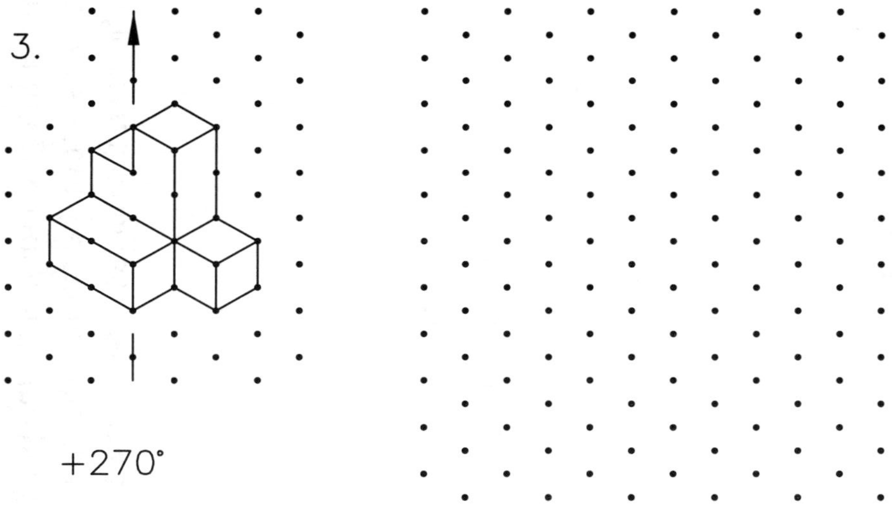

+270°

4.

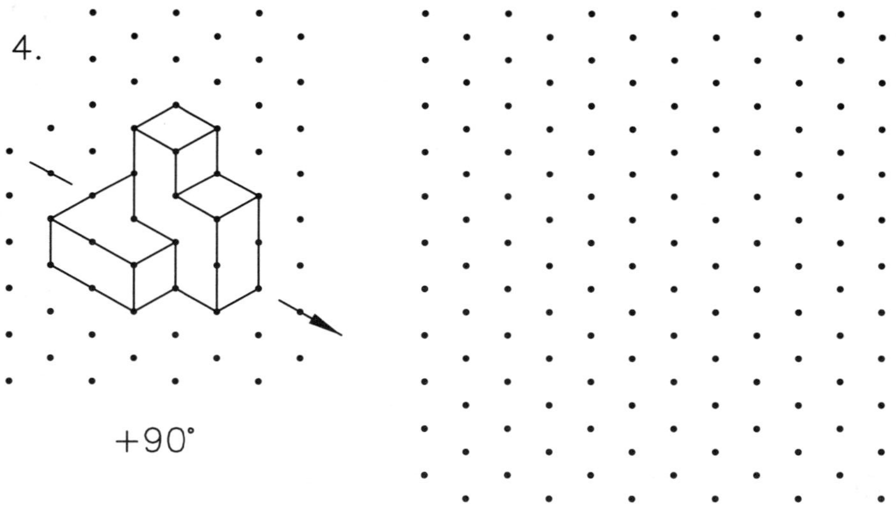

+90°

Name:

Class:

Section:

Date:

Developing Spatial Thinking

Grade:

Page

rot1−01

For the objects shown below sketch the object in the space provided after rotating it about the axis by the indicated amount.

1.

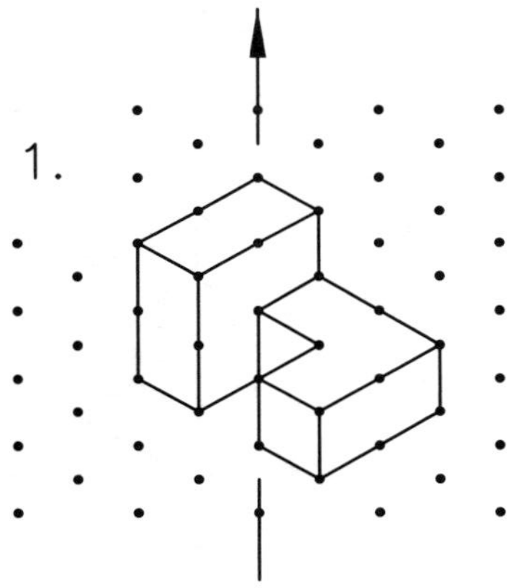

−90°

2.

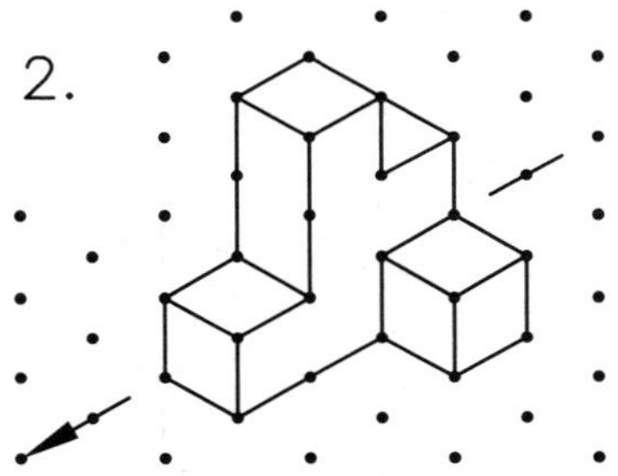

+90°

3.

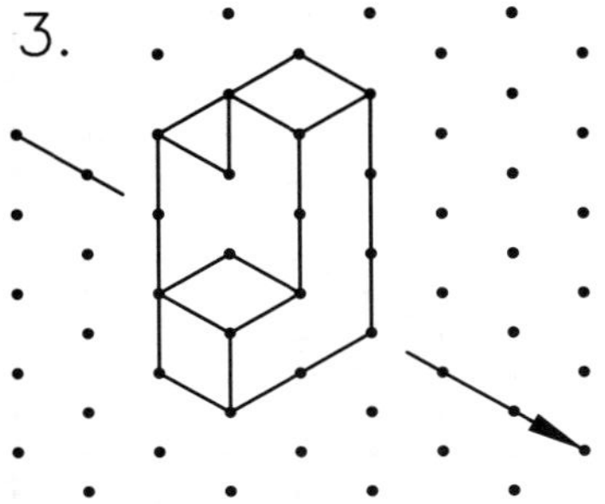

+180°

4.

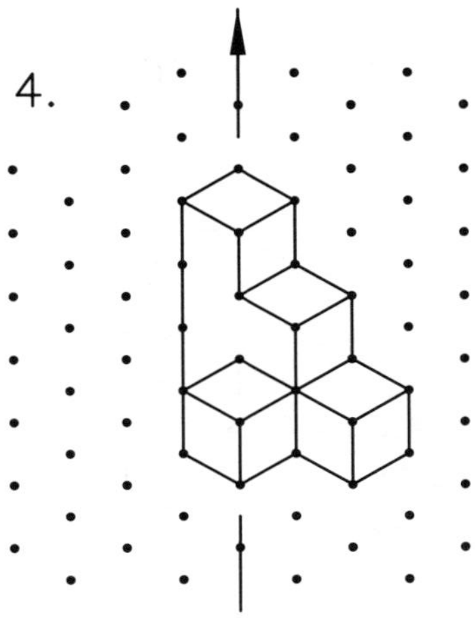

+180°

Name: Section: Date:

Class:

Developing Spatial Thinking   Grade: Page
                                          rot1−02

For the objects shown below sketch the object in the space provided after rotating it about the axis by the indicated amount.

1.

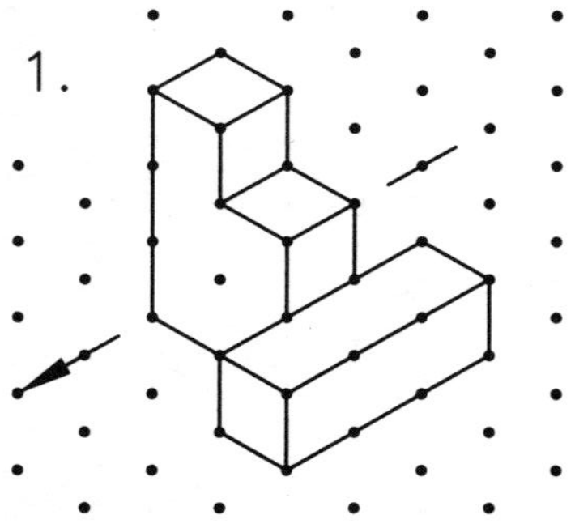

+180°

2.

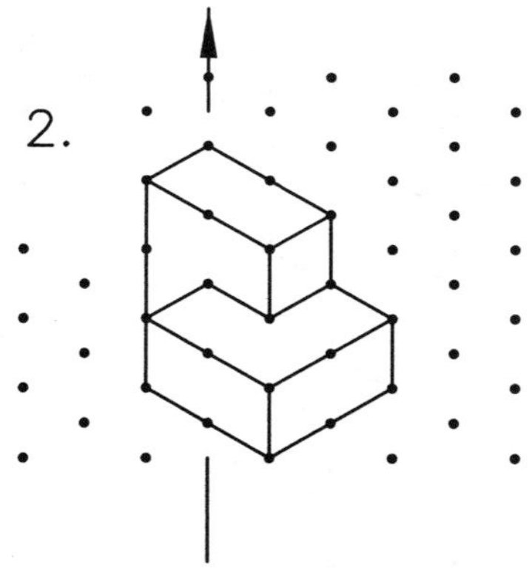

+90°

3.

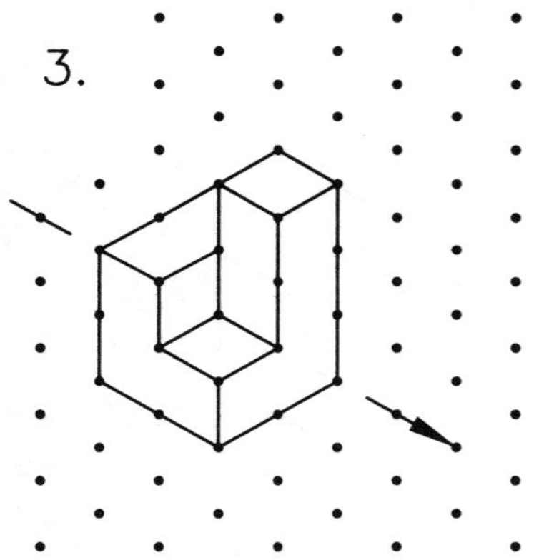

+270°

4.

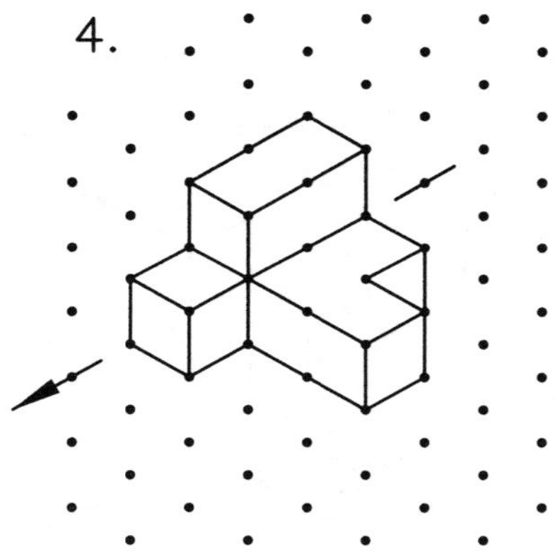

−180°

Name:
Class:

Section:
Date:

Developing Spatial Thinking

Grade:

Page
rot1—03

For the objects shown below sketch the object in the space provided after rotating it about the axis by the indicated amount.

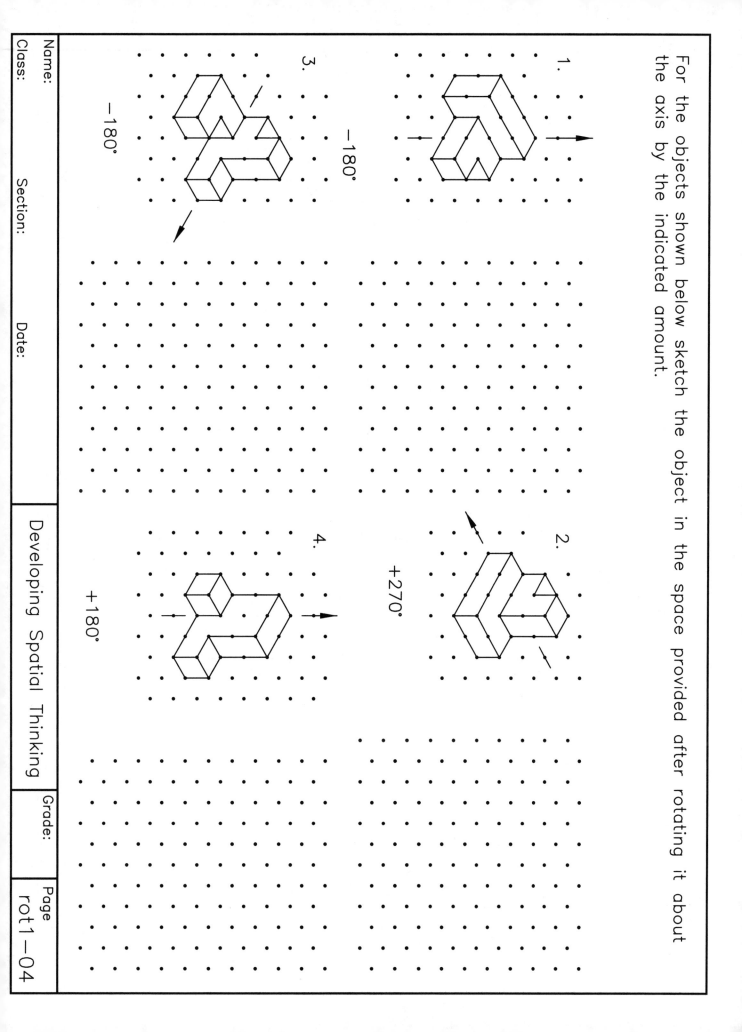

1.

2.

+270°

3.

−180°

4.

+180°

Name:
Class:

Section:

Date:

Developing Spatial Thinking

Grade:

Page
rot1−04

The objects shown below have been rotated positively about the given axis. In the space provided, indicate the amount of rotation (either 90°, 180°, or 270°).

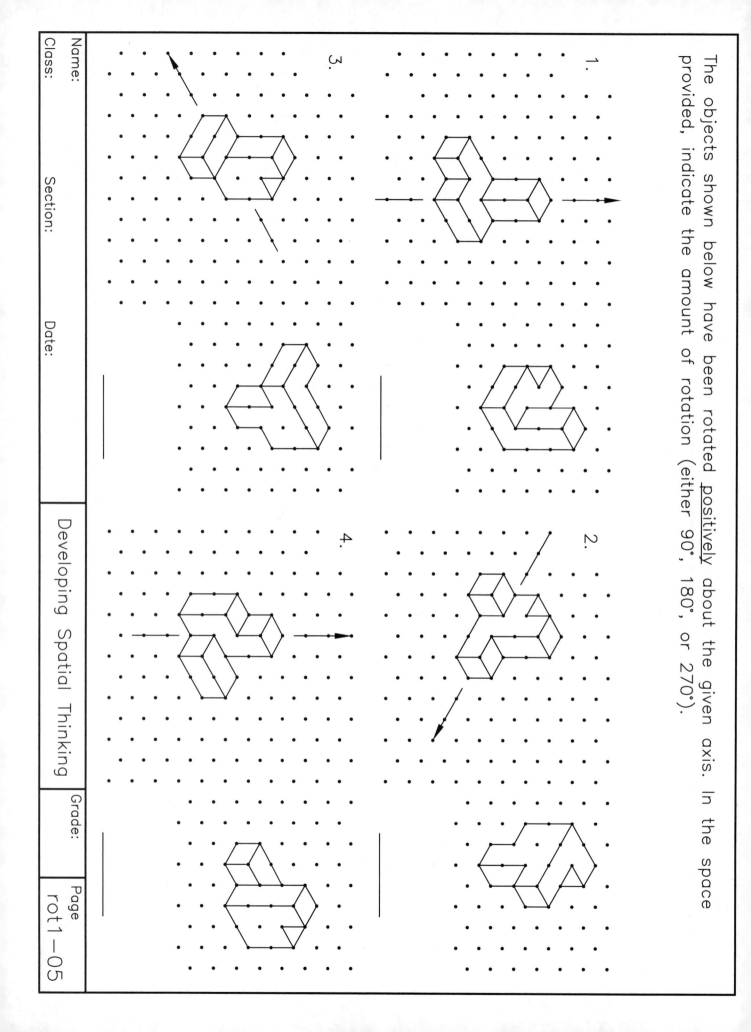

1.

2.

3.

4.

Name:
Class:

Section:

Date:

Developing Spatial Thinking

Grade:

Page
rot1—05

The objects shown below have been rotated _positively_ about the given axis. In the space provided, indicate the amount of rotation (either 90°, 180°, or 270°).

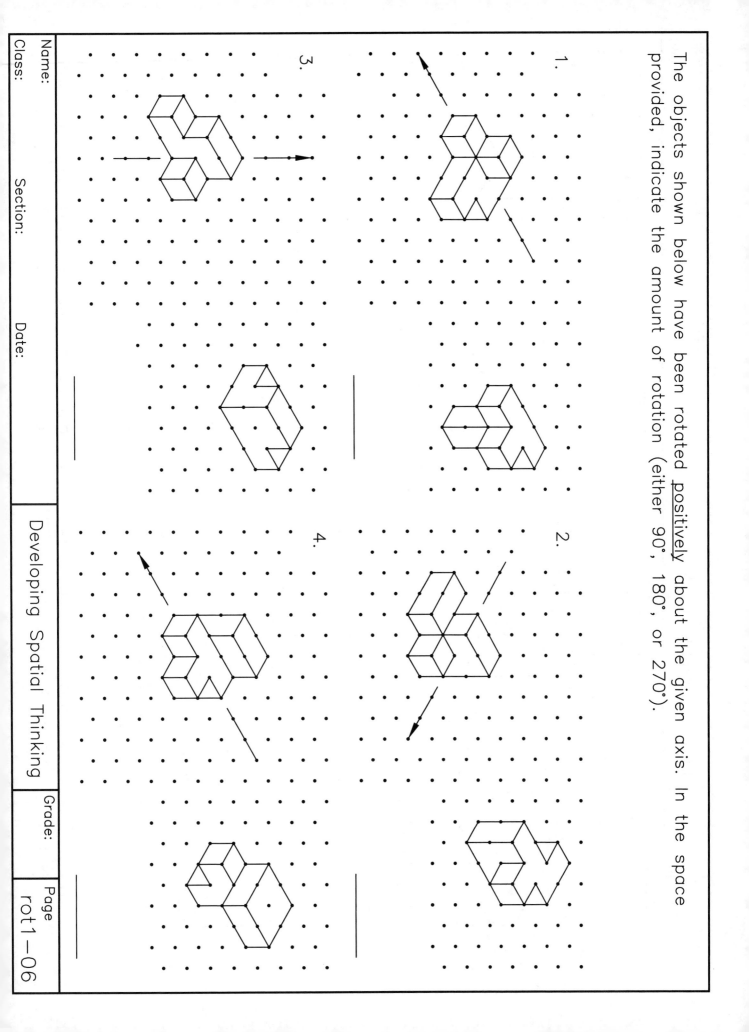

1.

2.

3.

4.

Name:

Class:

Section:

Date:

Developing Spatial Thinking

The objects shown below have been rotated _negatively_ about the given axis. In the space provided, indicate the amount of rotation (either 90°, 180°, or 270°).

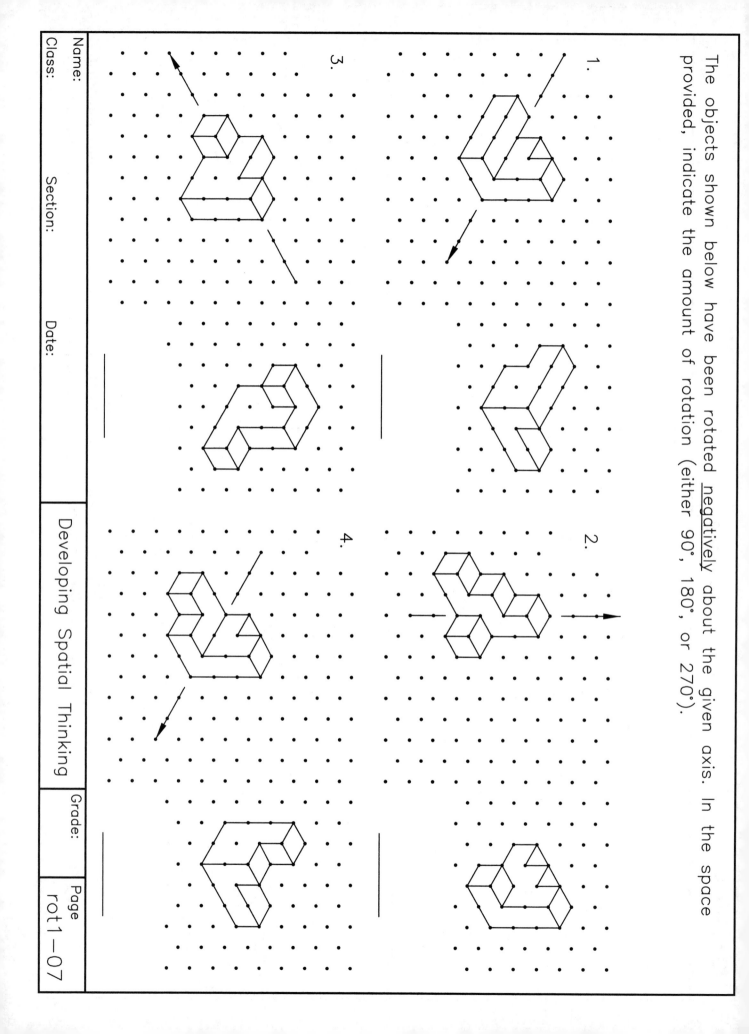

1.

2.

3.

4.

Name:

Class:

Section:

Date:

Developing Spatial Thinking

Grade:

Page
rot1—07

The objects shown below have been rotated _negatively_ about the given axis. In the space provided, indicate the amount of rotation (either 90°, 180°, or 270°).

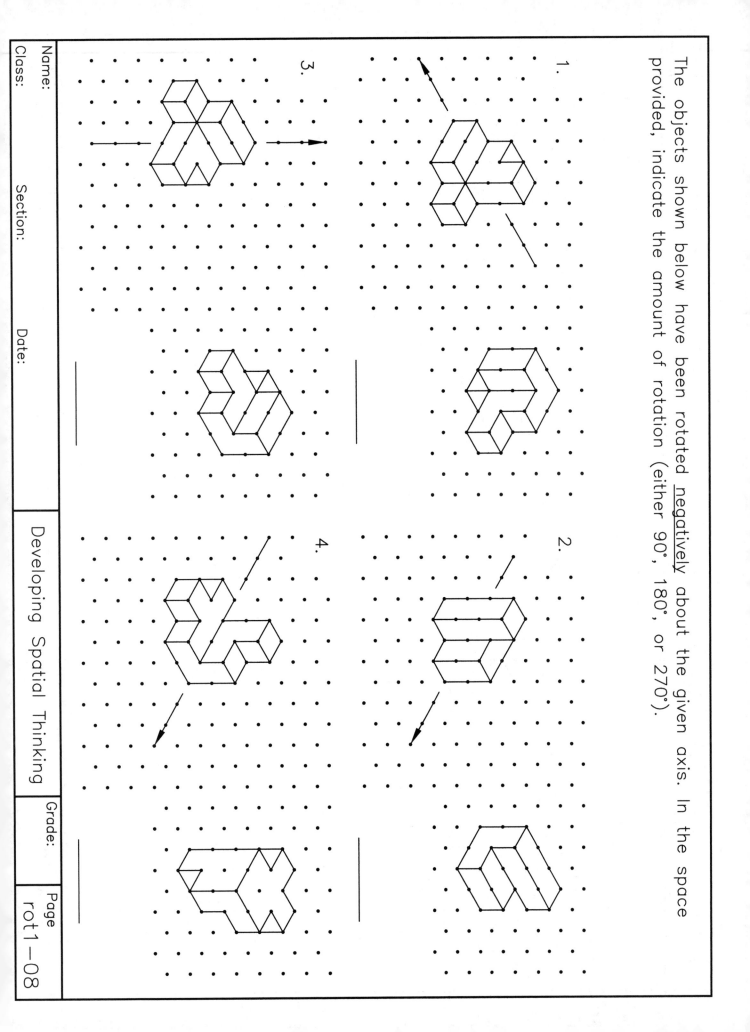

1.

2.

3.

4.

Name:
Class:

Section:
Date:

Developing Spatial Thinking

Grade:

Page
rot1—08

The objects shown below have been rotated _positively_ about one of the axes shown. Indicate the amount of rotation by inserting the appropriate rotation code in the space provided.

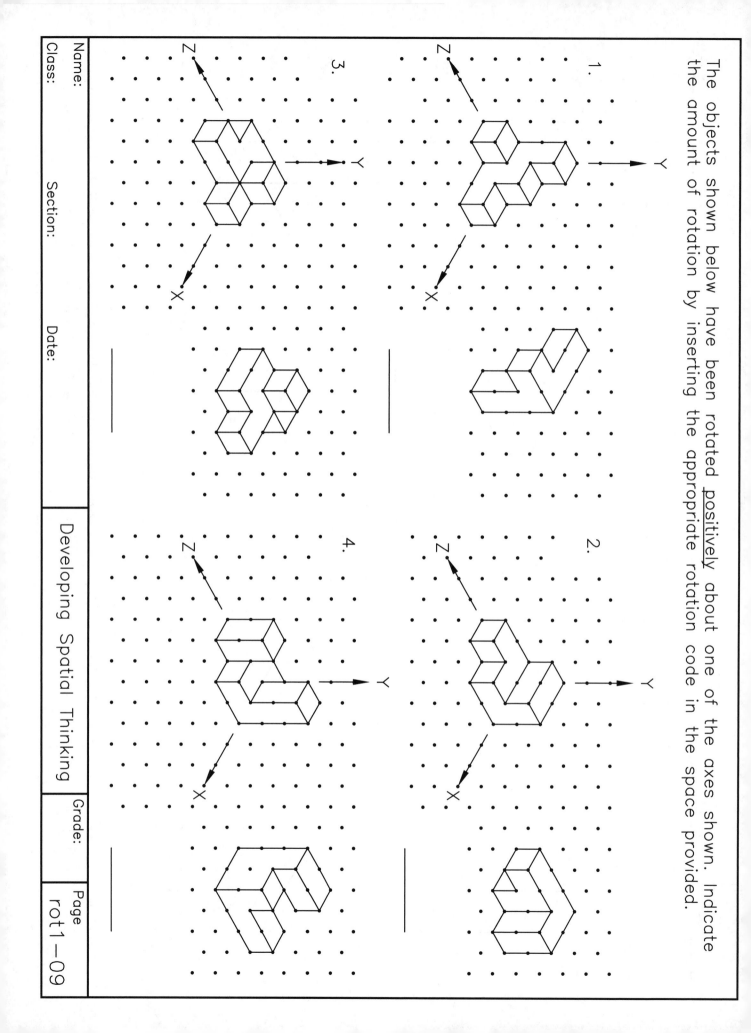

The objects shown below have been rotated _positively_ about one of the axes shown. Indicate the amount of rotation by inserting the appropriate rotation code in the space provided.

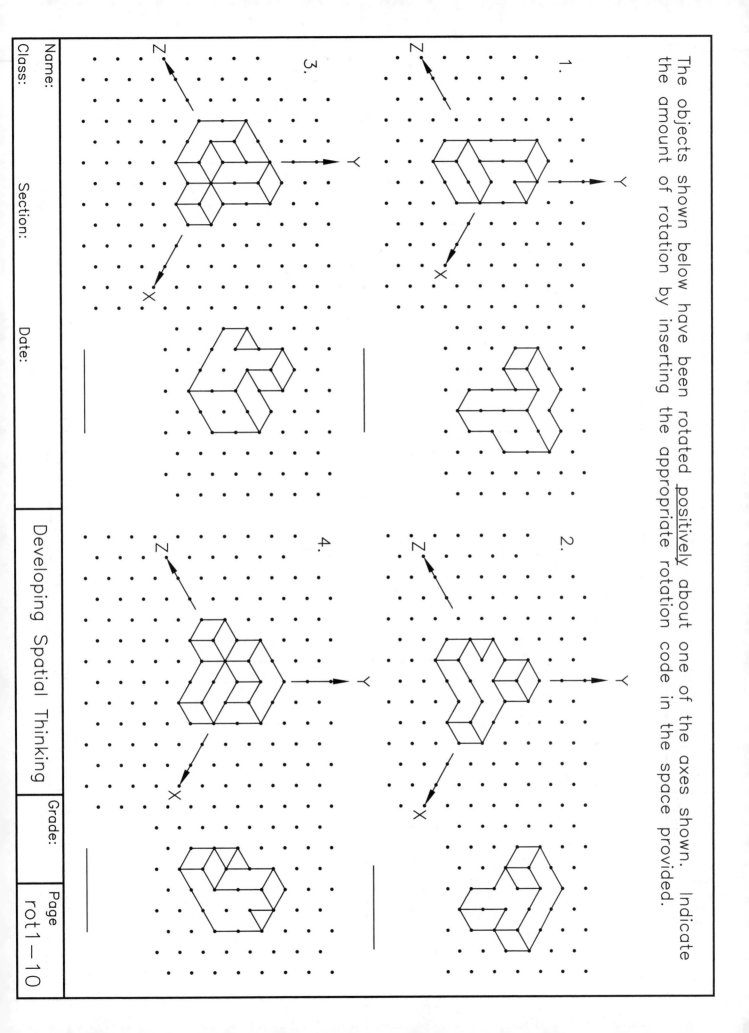

1.

2.

3.

4.

Name:

Class:

Section:

Date:

Developing Spatial Thinking

Grade:

Page
rot1—10

The objects shown below have been rotated positively about one of the axes shown. Indicate the amount of rotation by inserting the appropriate rotation code in the space provided.

1.

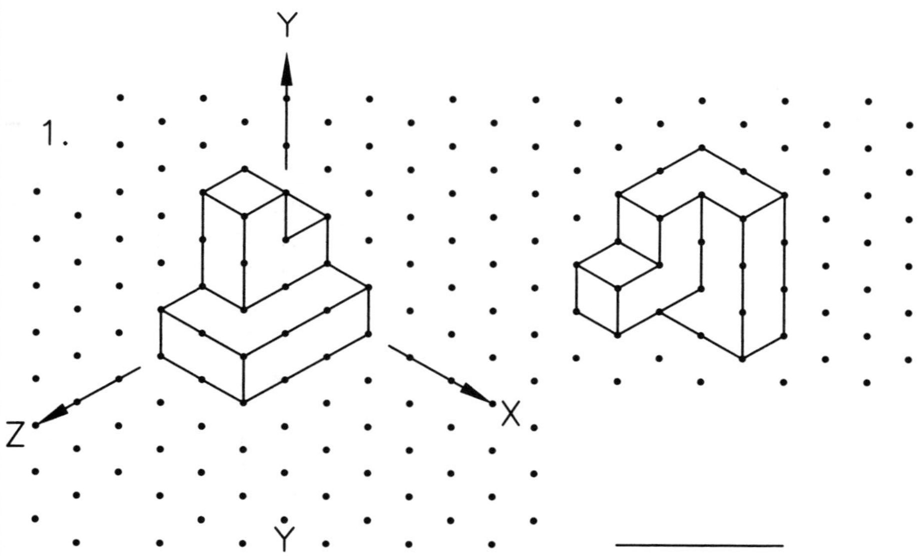

2.

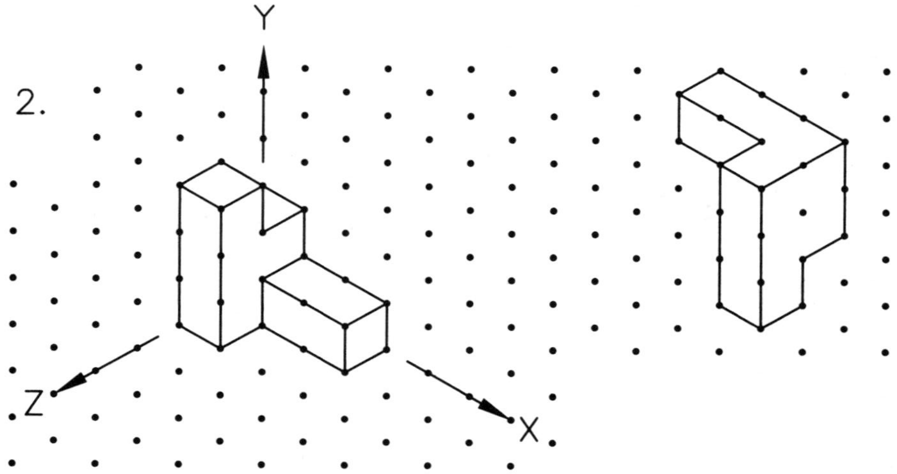

3.

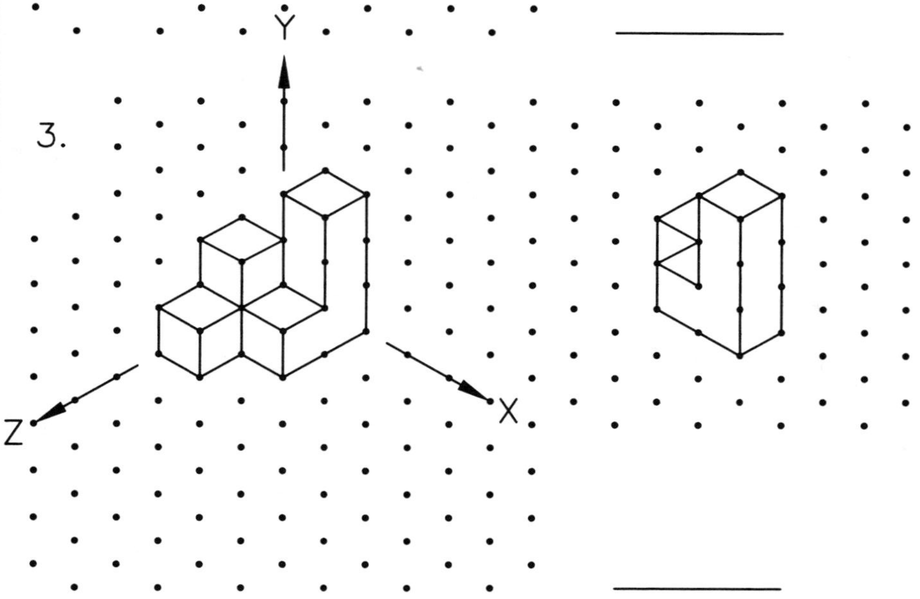

4.

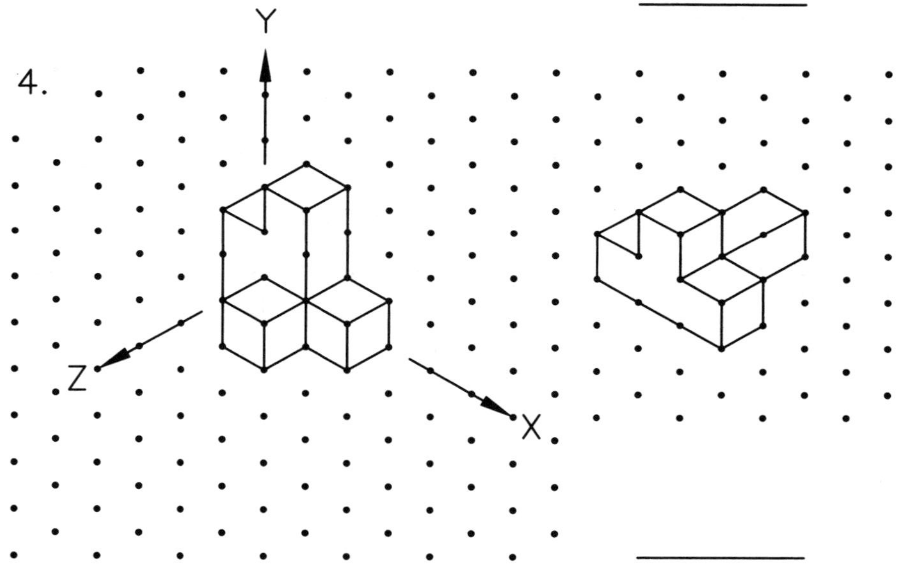

Name:
Class:

Section:

Date:

Developing Spatial Thinking

Grade:

Page
rot1−11

The objects shown below have been rotated positively about one of the axes shown. Indicate the amount of rotation by inserting the appropriate rotation code in the space provided.

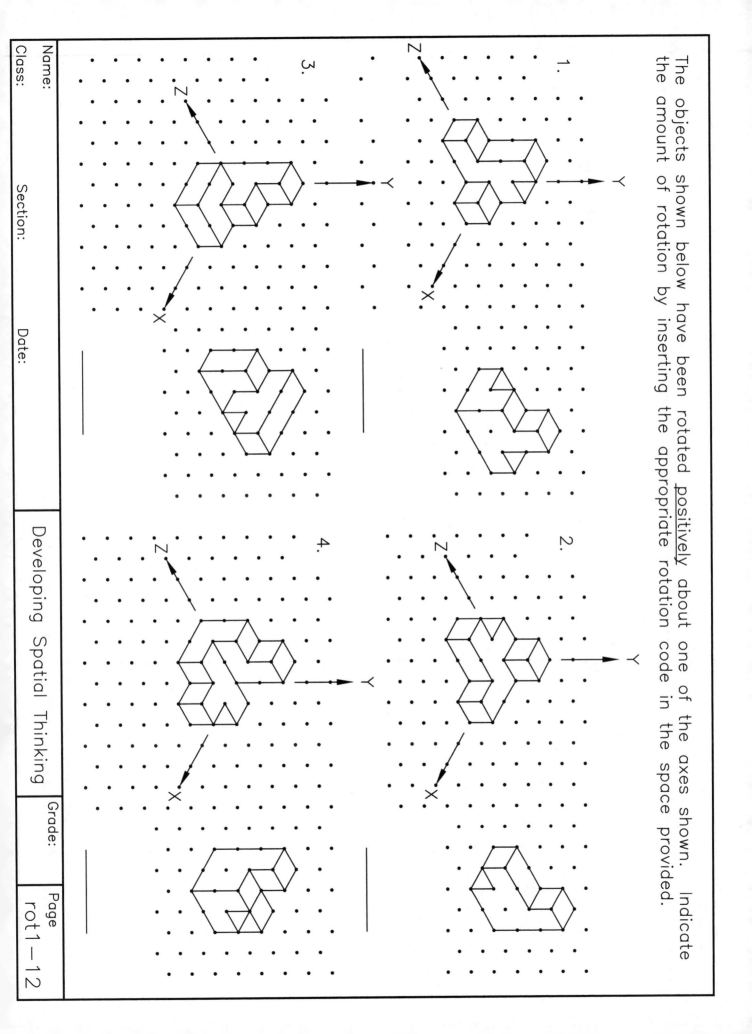

1.

2.

3.

4.

Name:

Class:

Section:

Date:

Developing Spatial Thinking

Grade:

The objects shown below have experienced the rotations given by the arrow codings. In the space provided, indicate an equivalent rotation code that would produce the same image.

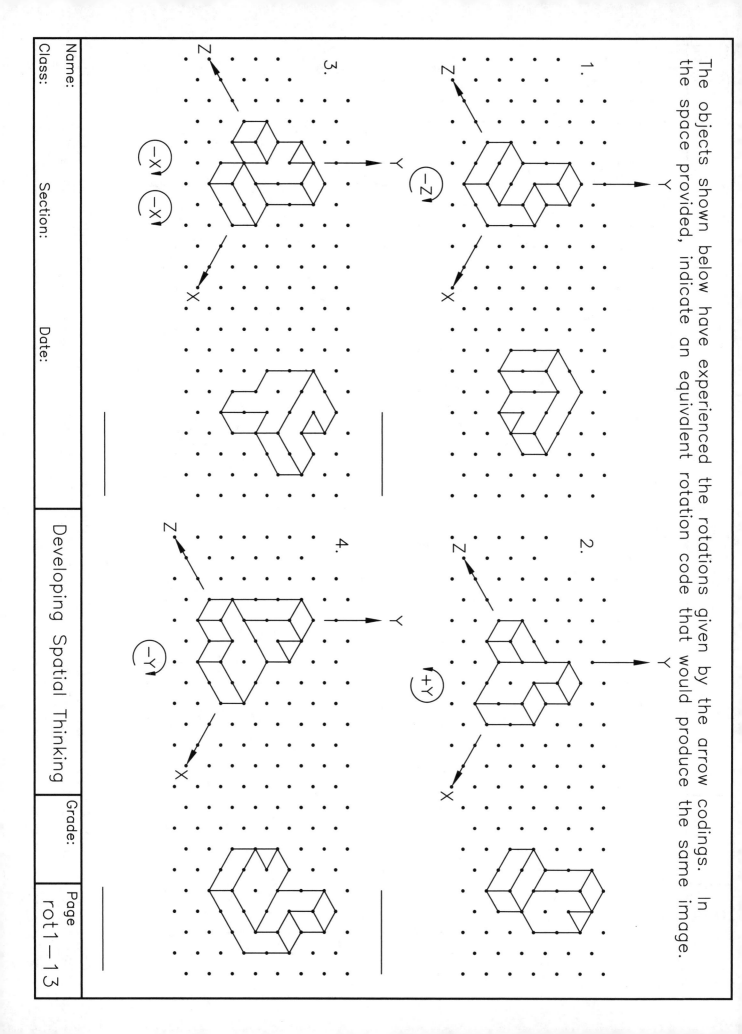

1.

2.

3.

4.

Name:
Class:

Section:

Date:

Developing Spatial Thinking

Grade:

Page
rot1—13

The objects shown below have experienced the rotations given by the arrow codings. In the space provided, indicate an equivalent rotation code that would produce the same image.

1.

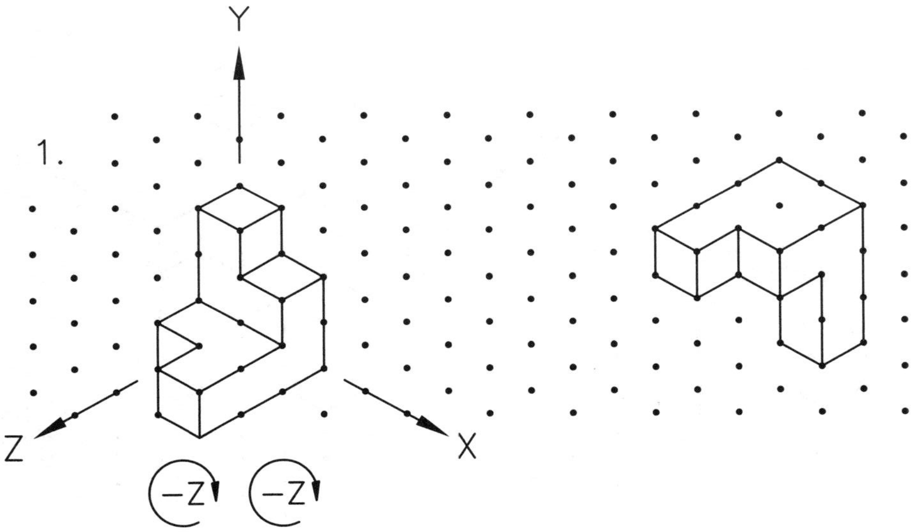

-Z
-Z

2.

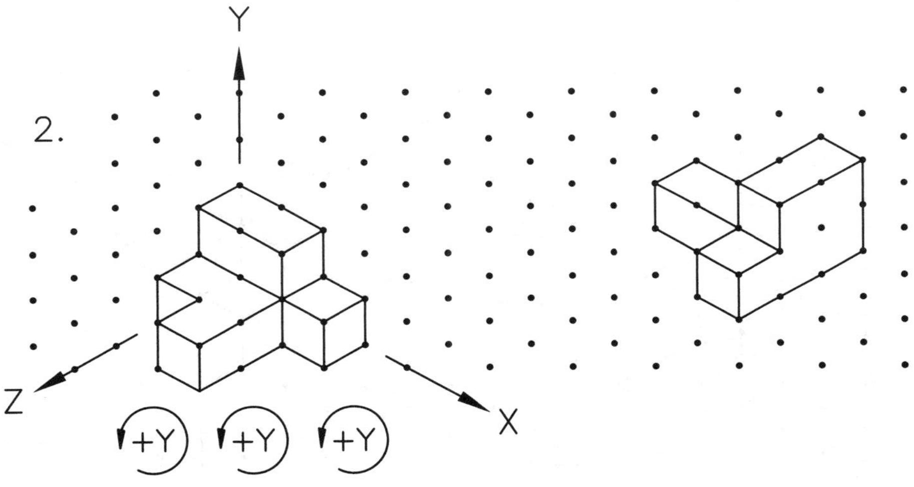

+Y
+Y
+Y

3.

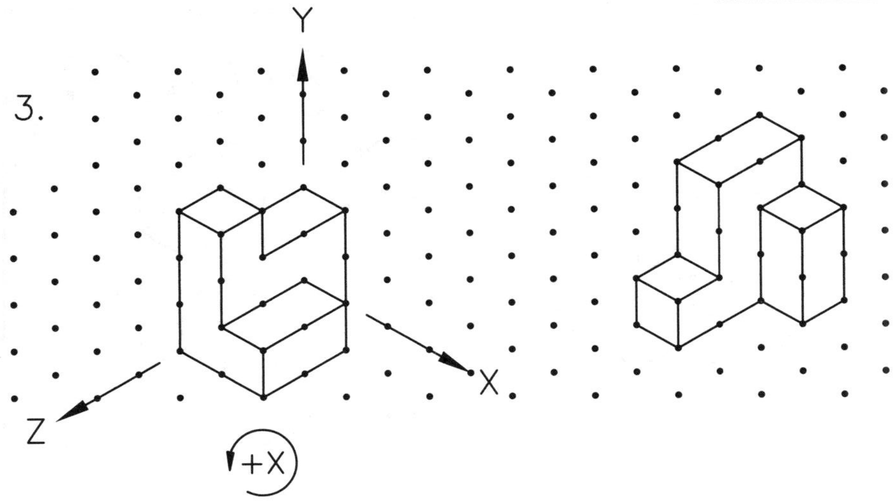

+X

4.

+Z

Name:
Class:

Section:

Date:

Developing Spatial Thinking

Grade:

Page
rot1 — 14

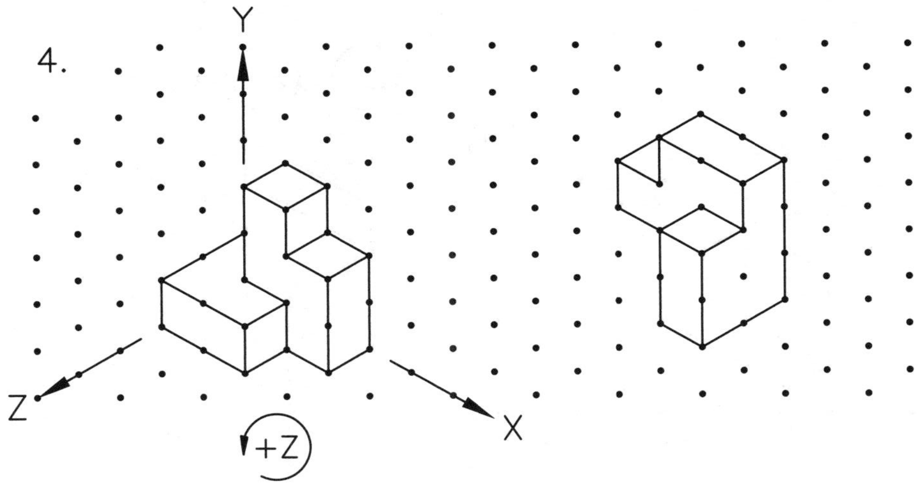

The objects shown below have experienced the rotations given by the arrows codings. In the space provided, indicate an equivalent rotation code that would produce the same image.

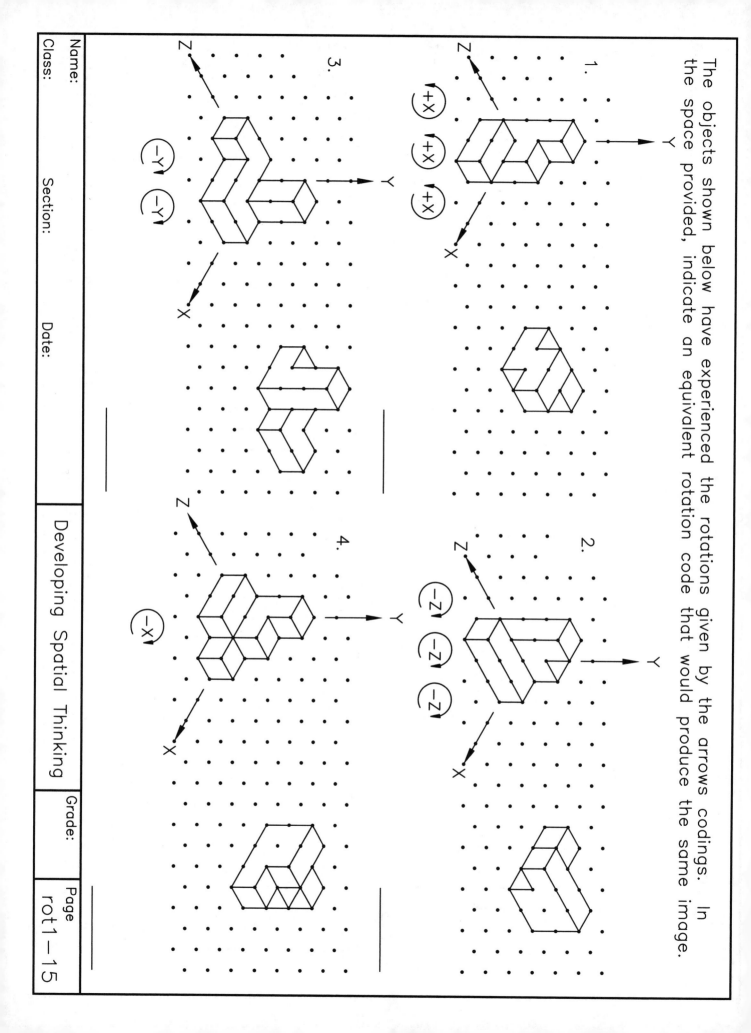

Name:

Class:

Section:

Date:

Developing Spatial Thinking

Grade:

Page
rot1—15

Rotate the objects shown below by the indicated amount and sketch the result in the space provided. You do not need to include the coordinate axes in your sketch.

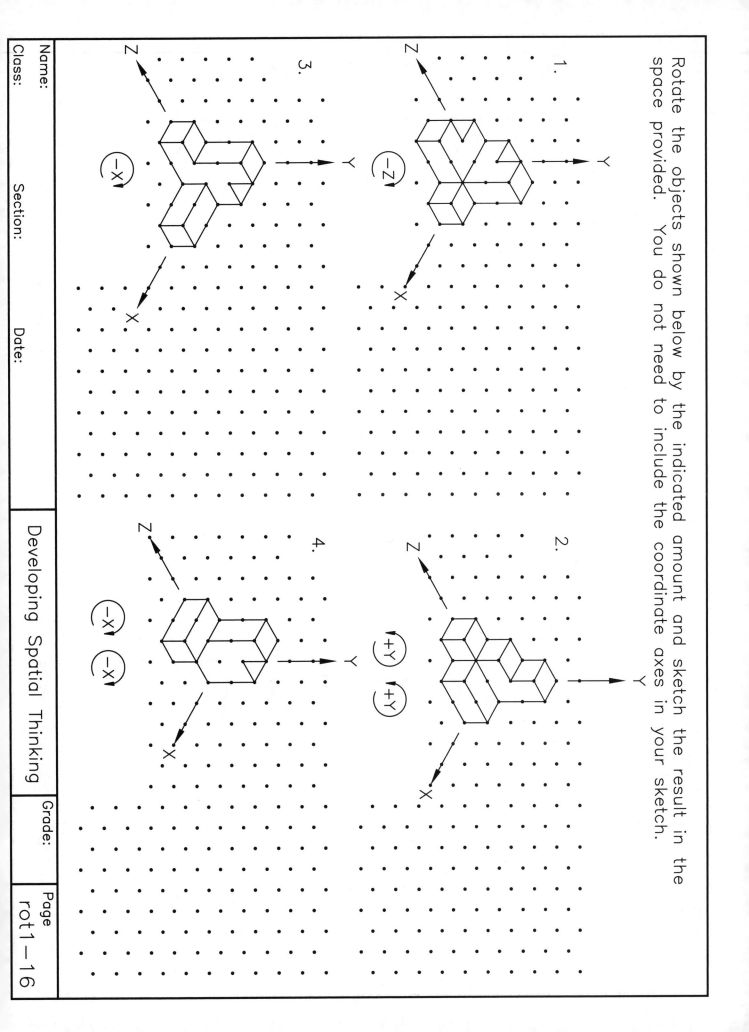

1.

2.

3.

4.

Name:
Class:

Section:

Date:

Developing Spatial Thinking

Grade:

Page
rot1—16

Rotate the objects shown below by the indicated amount and sketch the result in the space provided. You do not need to include the coordinate axes in your sketch.

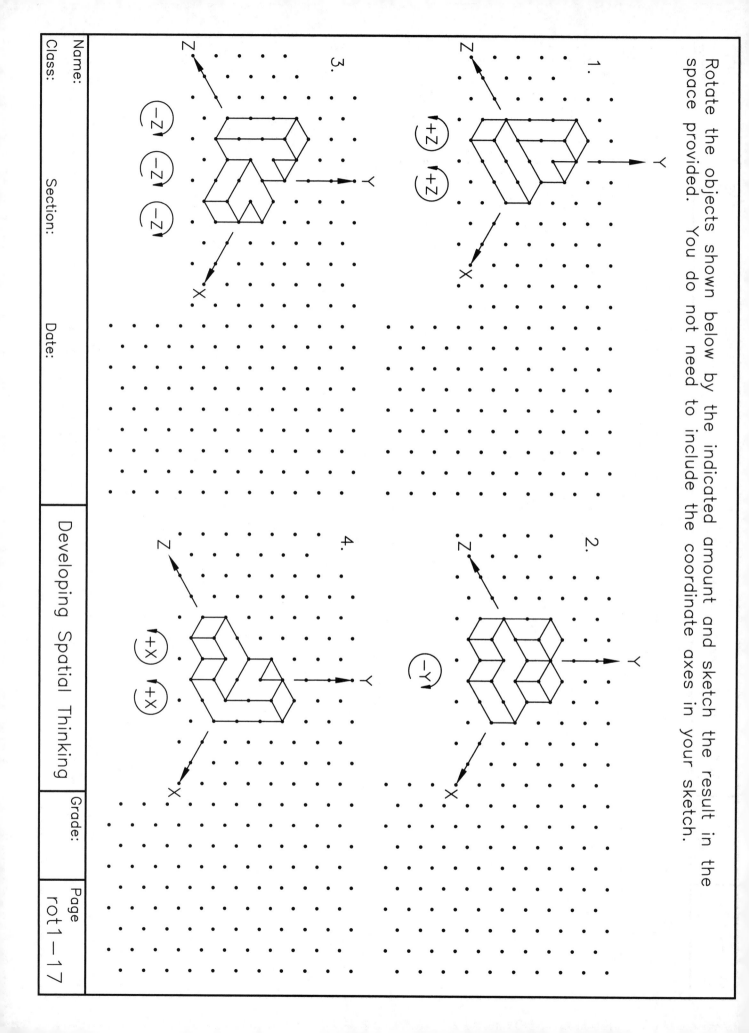

1.

2.

3.

4.

Name:
Class:

Section:

Date:

Developing Spatial Thinking

Grade:

Page
rot1 – 17

Rotate the objects shown below by the indicated amount and sketch the result in the space provided. You do not need to include the coordinate axes in your sketch.

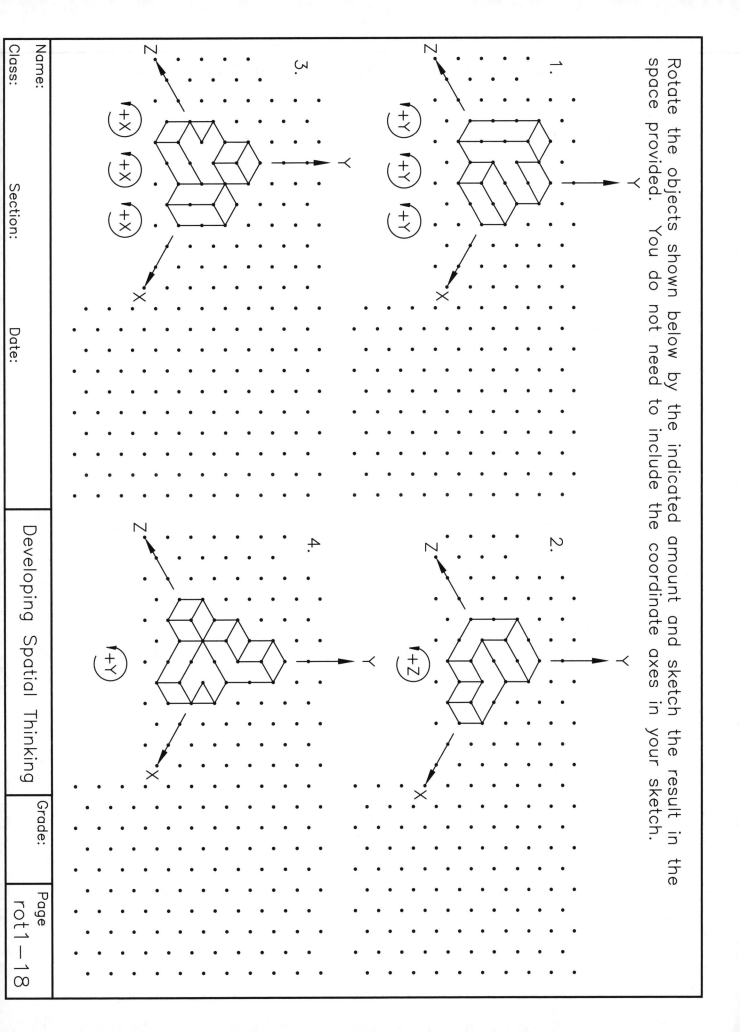

1.

2.

3.

4.

Name:

Class:

Section:

Date:

Rotation of Objects about Two or More Axes

Objects can be rotated about two or more axes in the same manner that they can be rotated about a single axis.

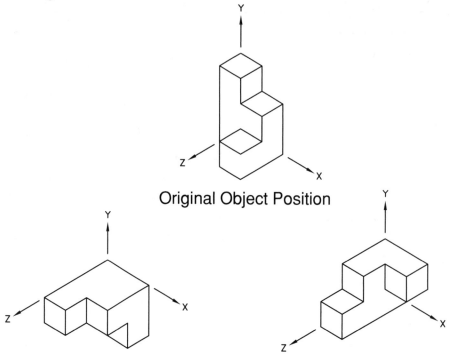

Original Object Position

Object First Rotated about Positive X

Then Rotated about Positive Z

When rotation occurs about a single axis, an entire edge remains in its original position. When rotation occurs about two different axes, only a single pivot point remains in its original position.

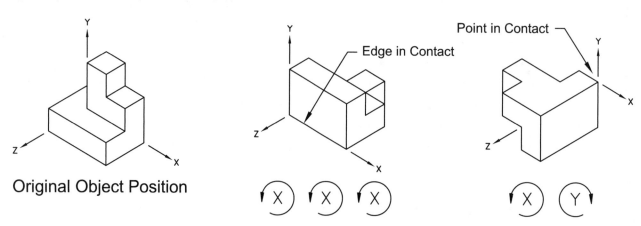

Original Object Position

Edge in Contact

Point in Contact

Objects can be positively (counterclockwise) or negatively (clockwise) rotated about each axis, just like rotating about a single axis. (Remember the right hand rule!)

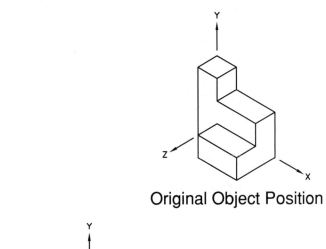

Original Object Position

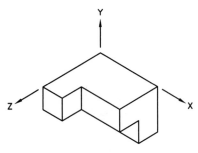

Object First Rotated about Positive X

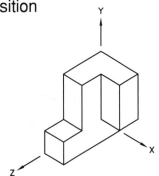

Then Rotated about Positive Z

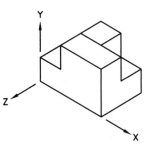

Object First Rotated about Negative X

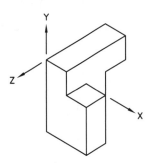

Then Rotated about Negative Z

Rotations about two or more axes are not commutative. In other words, the order in which the rotations occur is important. If you switch the order of the rotations, you will not end up with the same result.

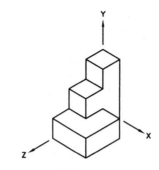

Original Object Position

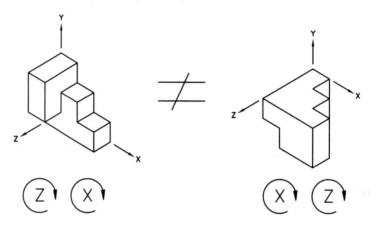

You can rotate an object in space as many times as you wish which can result in complex rotation designation codes.

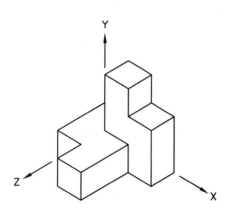

Original Object Position

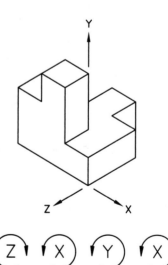

As with single axis rotations, different rotation designation codes can give you the same result. Two sets of rotation designation codes that result in the same final orientation of the object are said to be *equivalent*.

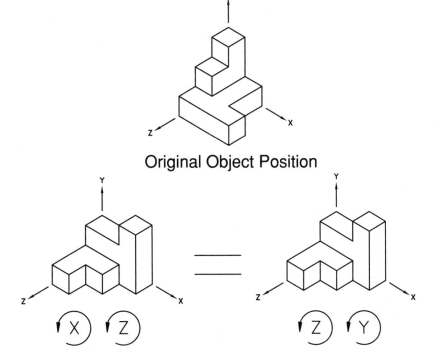

Complex rotation designation codes can sometimes be reduced to simpler ones.

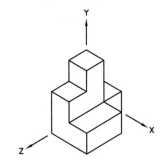

Original Object Position

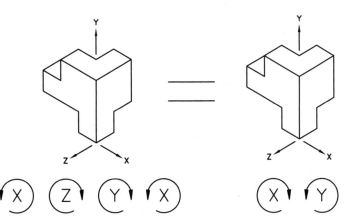

Circle the letter corresponding to the view on the right that shows the result of rotating the object on the left by the indicated amount.

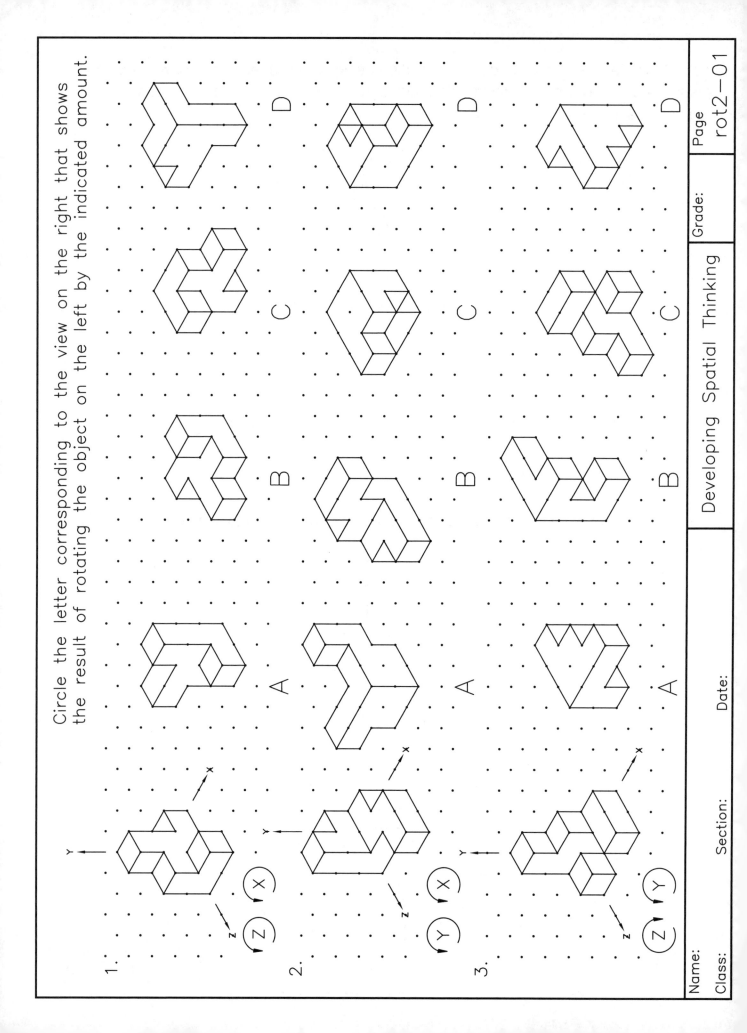

Name:

Class:

Section:

Date:

Developing Spatial Thinking

Grade:

Page
rot2−01

Circle the letter corresponding to the view on the right that shows the result of rotating the object on the left by the indicated amount.

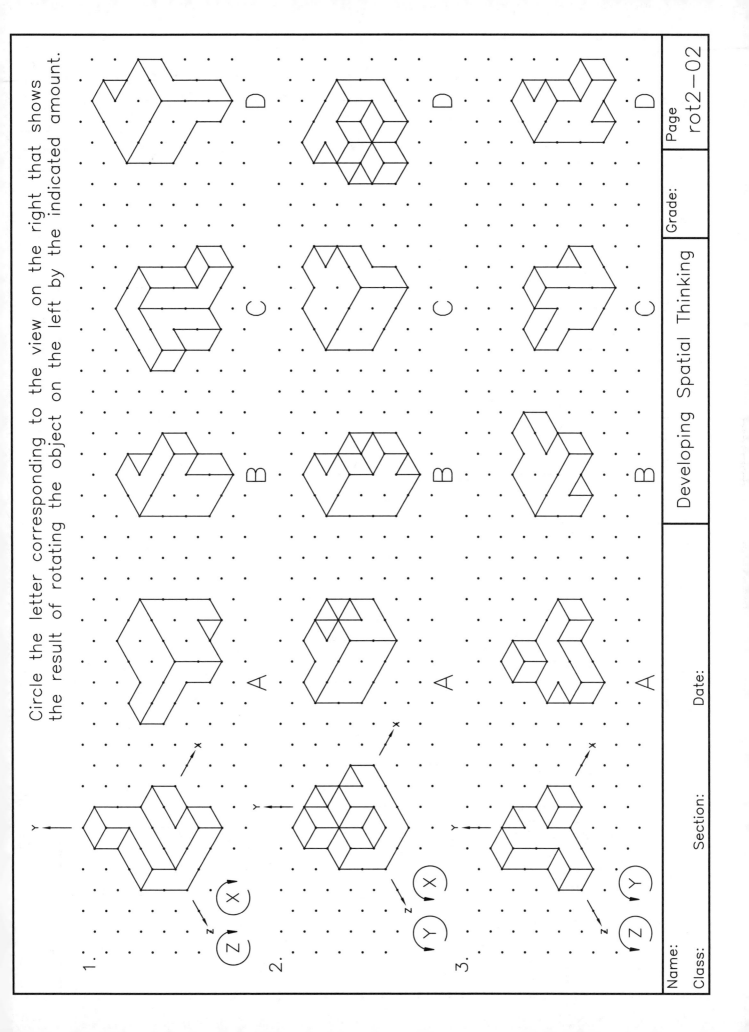

1. A   B   C   D

2. A   B   C   D

3. A   B   C   D

Developing Spatial Thinking

Name: Grade:

Class: Section: Date:

Circle the letter corresponding to the view on the right that shows the result of rotating the object on the left by the indicated amount.

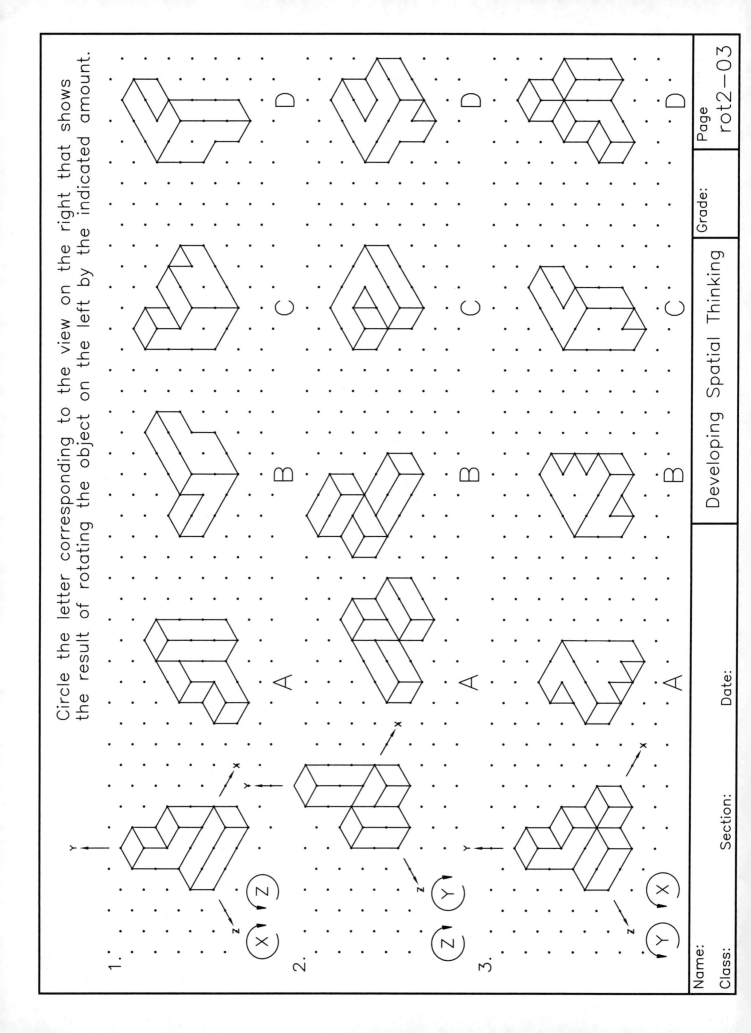

1.

A   B   C   D

2.

A   B   C   D

3.

A   B   C   D

Developing Spatial Thinking

Name:

Class:

Section:

Date:

Grade:

Circle the letter corresponding to the view on the right that shows the result of rotating the object on the left by the indicated amount.

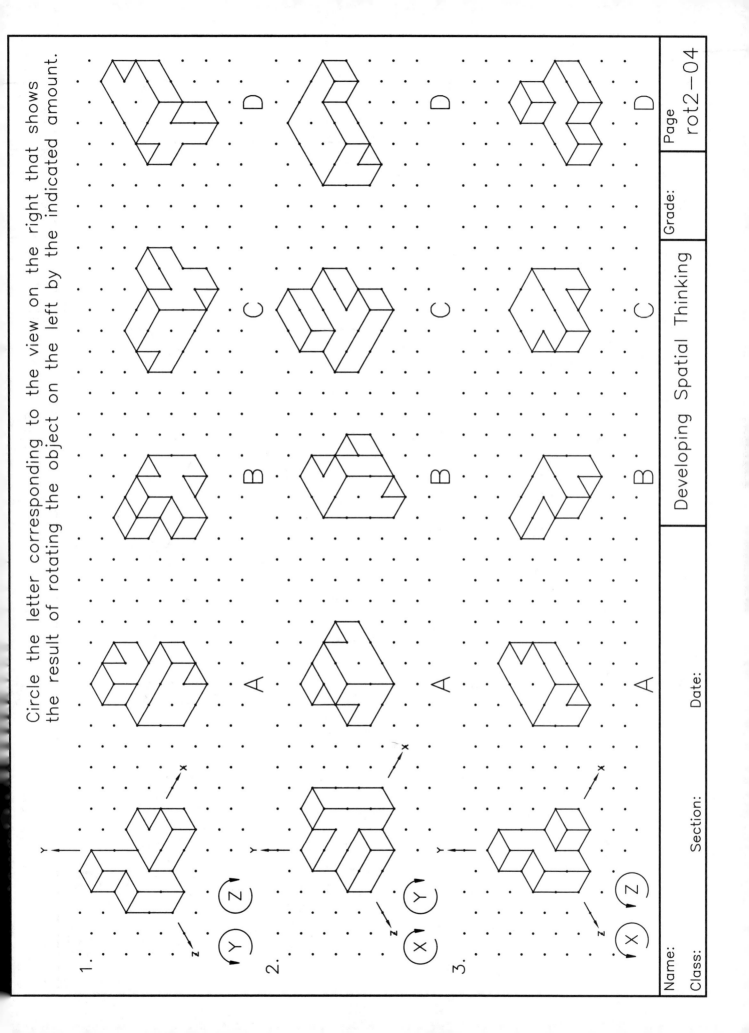

1.   Y   Z

A   B   C   D

2.   X   Y

A   B   C   D

3.   X   Z

A   B   C   D

Developing Spatial Thinking

Name:

Class: Section: Date:

Grade:

Page
rot2—04

Circle the letter corresponding to the view on the right that shows the result of rotating the object on the left by the indicated amount.

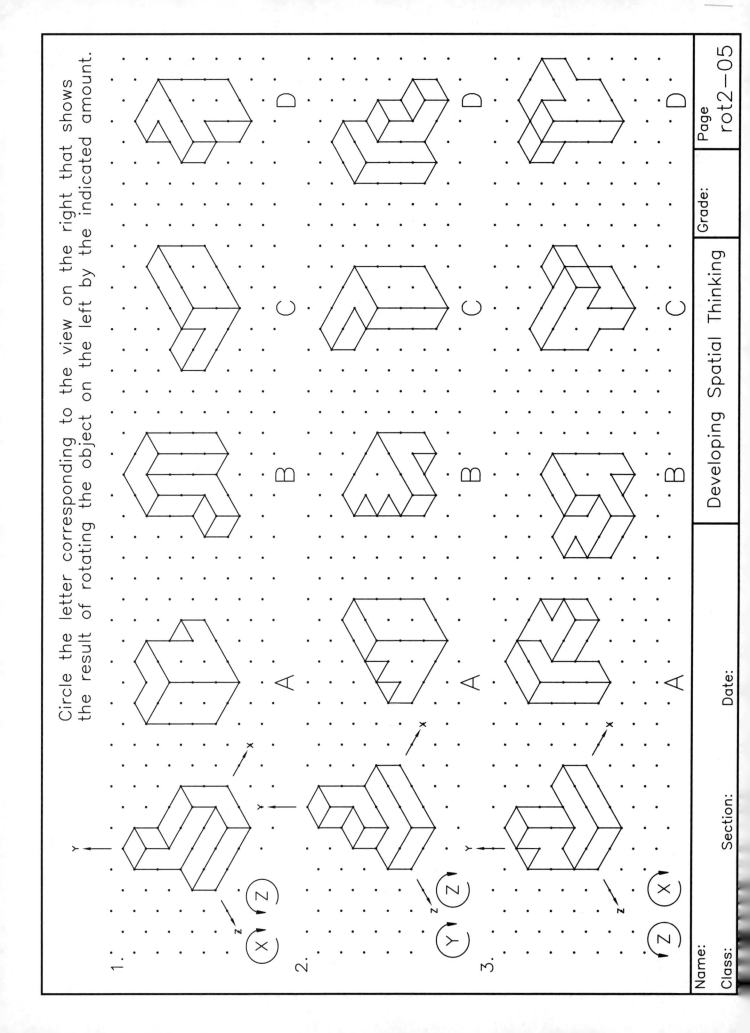

Developing Spatial Thinking

Grade:

Name:

Class:

Section:

Date:

Choose the rotation code that will rotate the object on the left to match the object on the right.

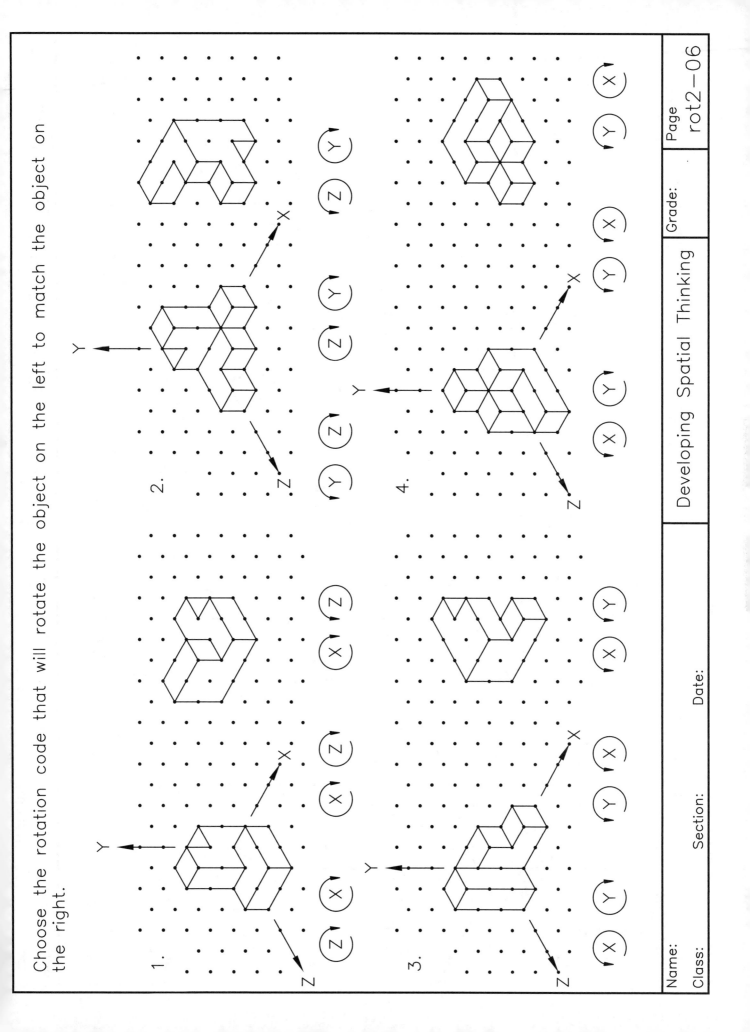

Developing Spatial Thinking

Name:

Class:

Section:

Grade:

Date:

Choose the rotation code that will rotate the object on the left to match the object on the right.

1.

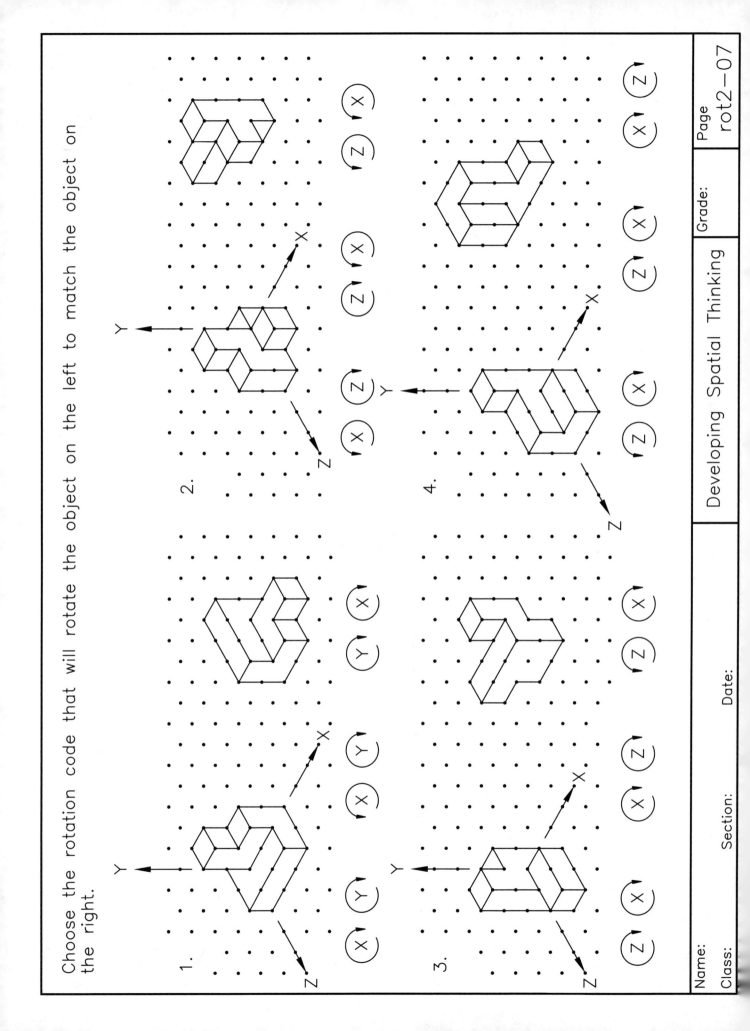

Name:

Class:

Section:

Date:

Developing Spatial Thinking

Grade:

Page
rot2-07

Choose the rotation code that will rotate the object on the left to match the object on the right.

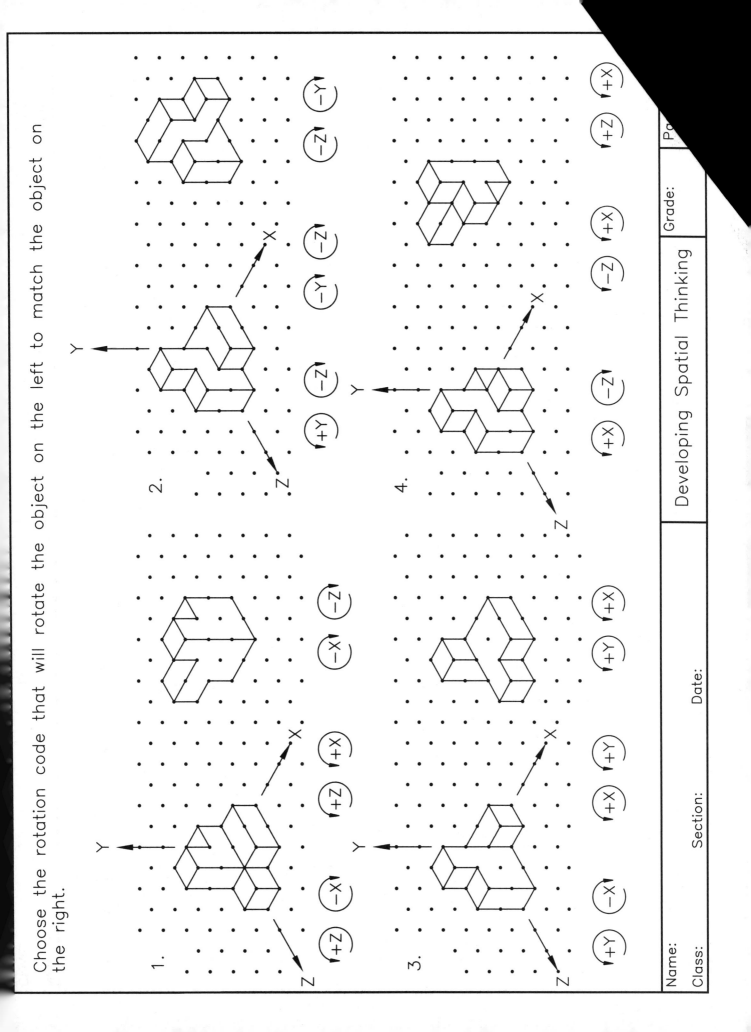

1. (+Z) (-X) (+X) (+Z) (-X) (-Z)

2. (+Y) (-Z) (-Y) (-Z) (-Z) (-Y)

3. (+Y) (+X) (+Y) (+Y) (+X) (+X)

4. (+X) (-Z) (-Z) (+X) (+Z) (+X)

Name: Section: Date:
Class: Developing Spatial Thinking   Grade: Pa

Choose the rotation code that will rotate the object on the left to match the object on the right.

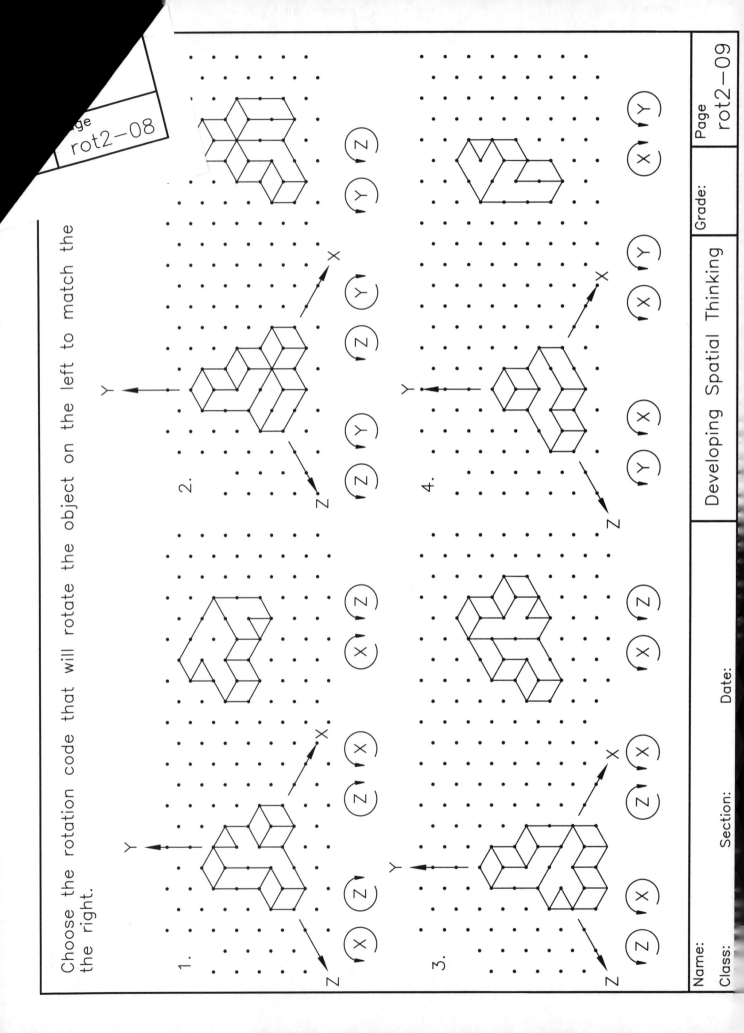

1.
Z → X   X → X   Z → X
Z → Z   Z → Z   X → X

2.
Z → Y   Z → Z   Y → Z
Z → Y   Y → Y   Z → Z

3.
Z → X   Z → X   X → X
Z → Z   X → X   Z → Z

4.
Y → Y   Y → X   X → Y
Y → X   X → Z   X → Y

In the space provided, indicate the rotation code that would rotate the object on the left to obtain the view of it shown on the right. (There may be more than one correct response.)

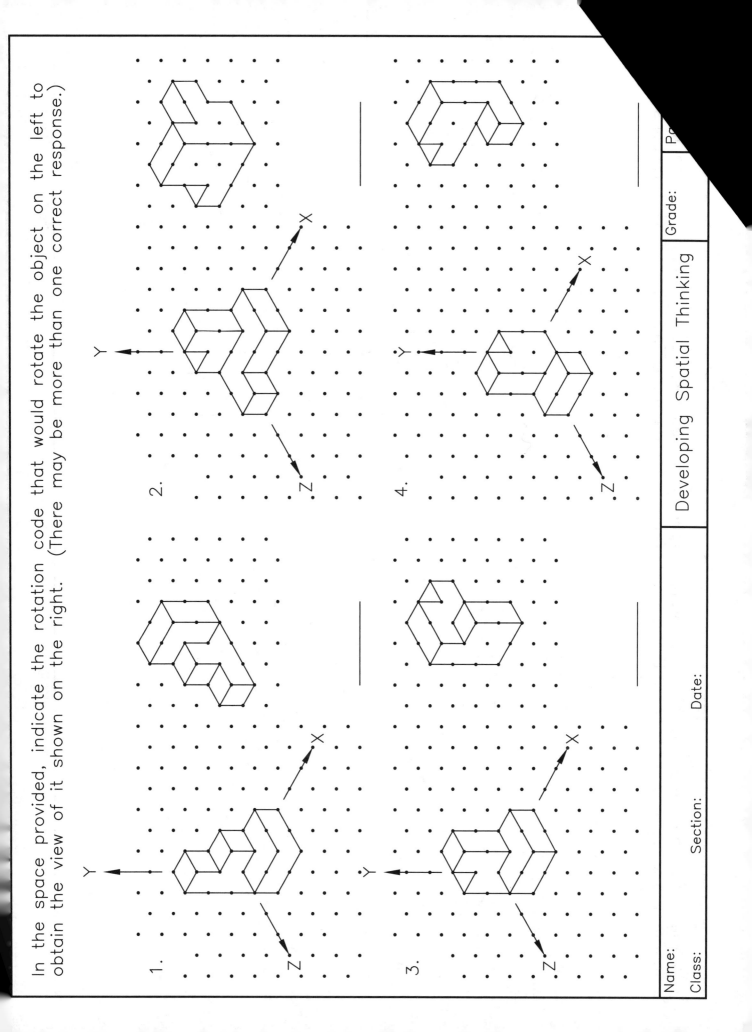

1.

2.

3.

4.

Developing Spatial Thinking

In the space provided, indicate the rotation code that would rotate the object ₍
obtain the view of it shown on the right. (There may be more than one corre

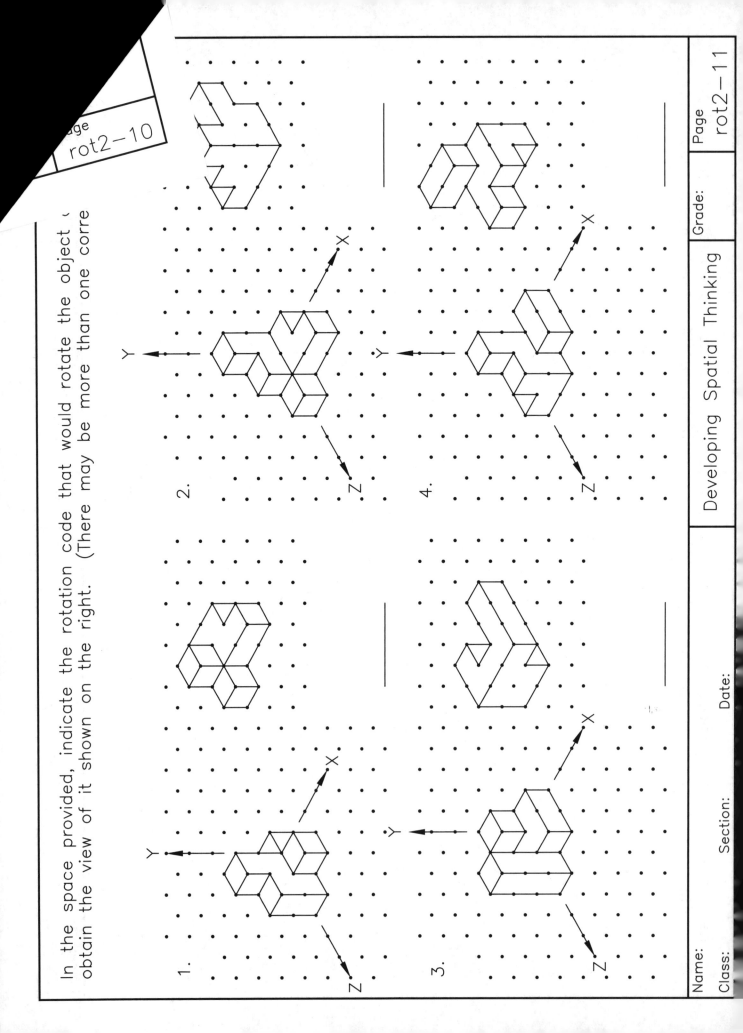

1.

2.

3.

4.

Name:

Class:

Section:

Date:

Developing Spatial Thinking

Grade:

Page
rot2-11

In the space provided, indicate the rotation code that would rotate the object on the left to obtain the view of it shown on the right. (There may be more than one correct response.)

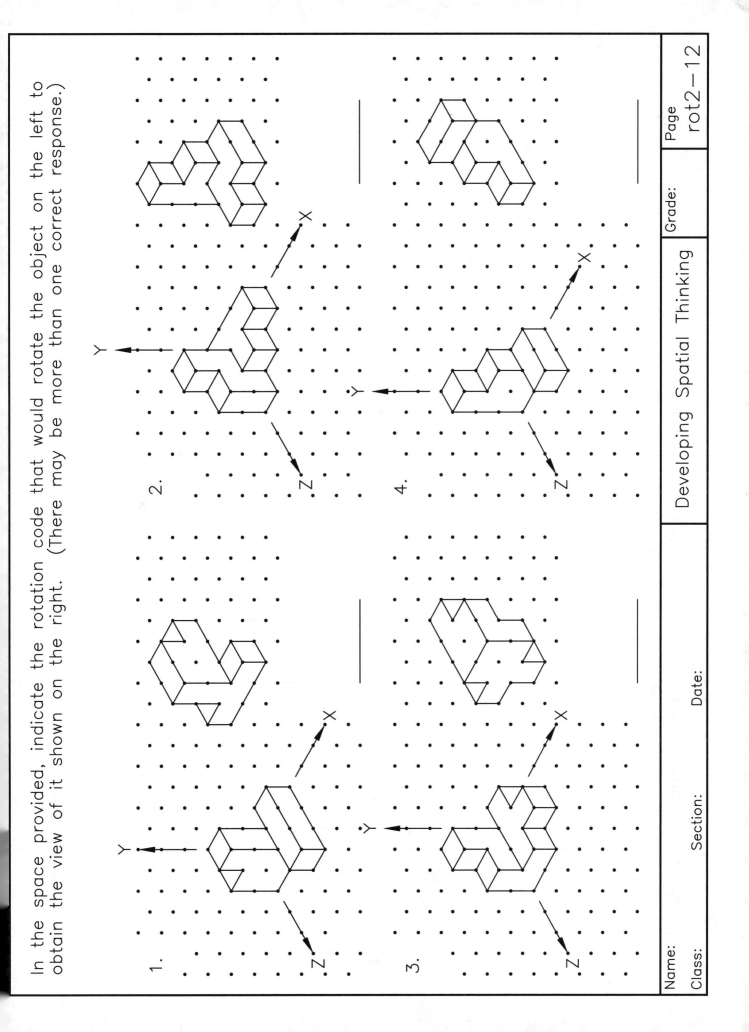

Name:

Class:

Section:

Date:

Developing Spatial Thinking

Grade:

Page

rot2−12

In the space provided, indicate the rotation code that would rotate the object on the left to obtain the view of it shown on the right. (There may be more than one correct response.)

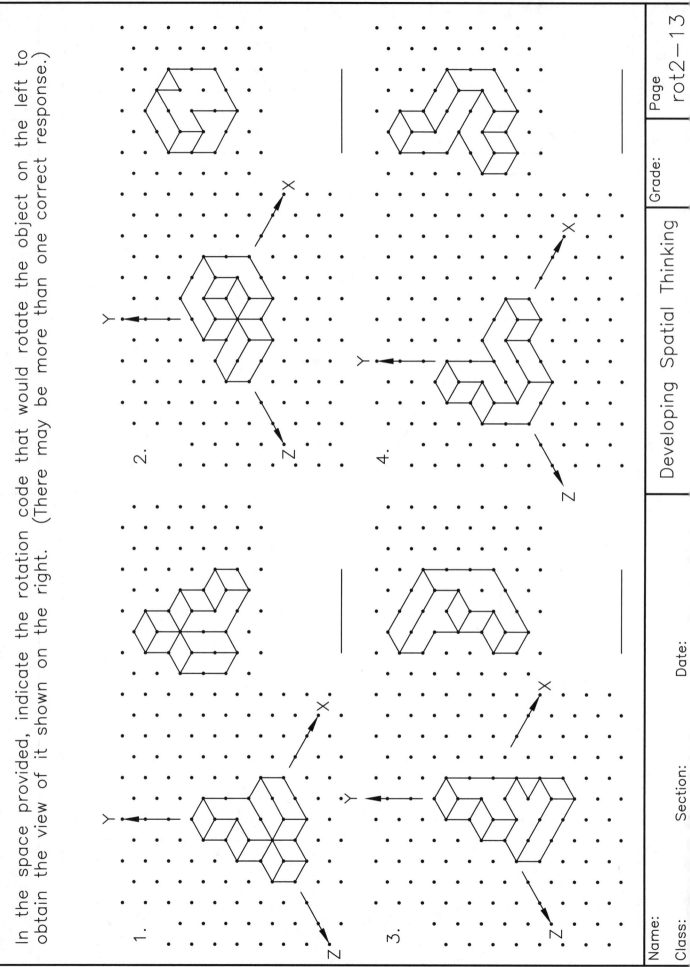

Name:

Class:

Section:

Date:

Developing Spatial Thinking

Grade:

Rotate the objects shown below by the indicated amount. Sketch the result in the space provided. Make sure you perform the rotations in the given order.

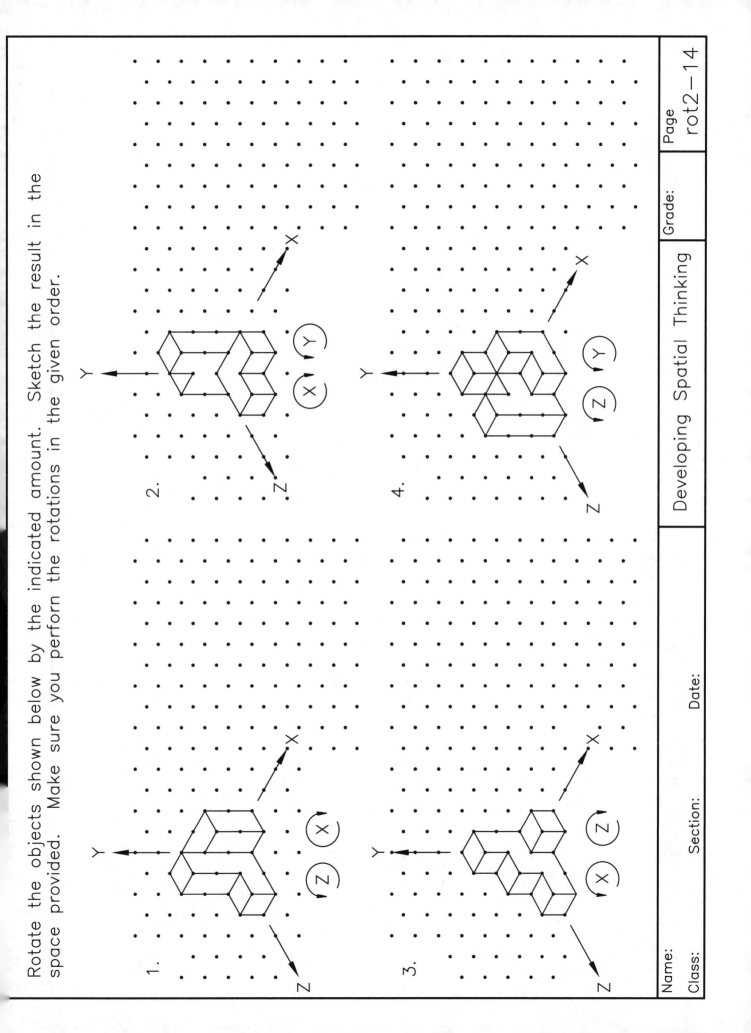

Name:

Class:

Section:

Date:

Developing Spatial Thinking

Grade:

Page
rot2—14

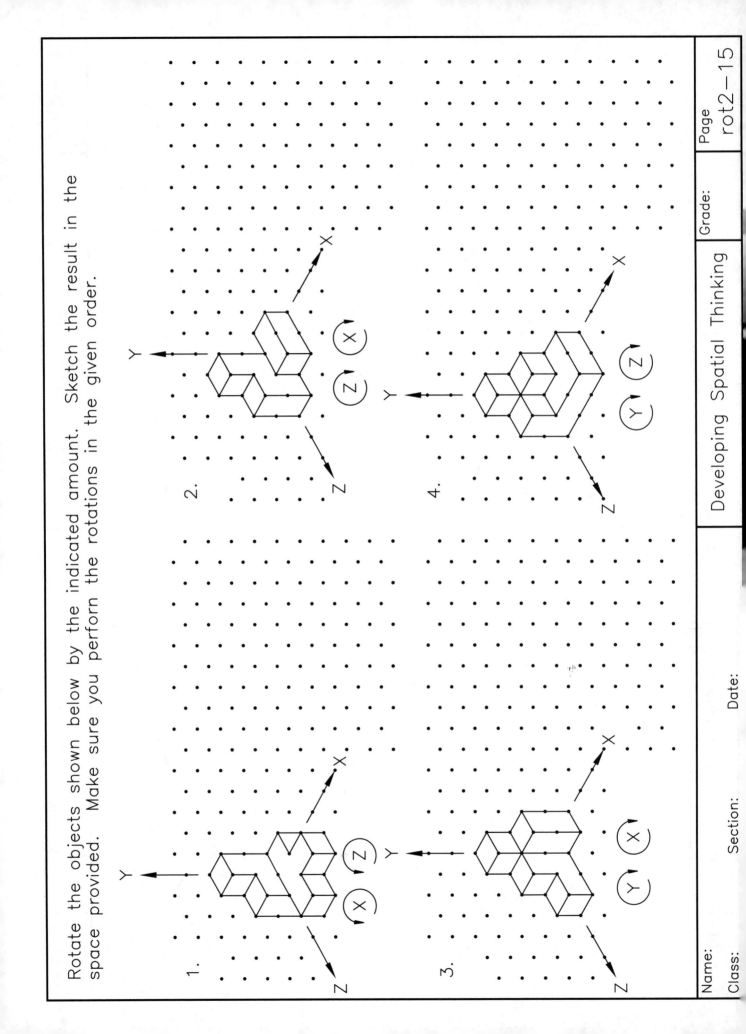

Rotate the objects shown below by the indicated amount. Sketch the result in the space provided. Make sure you perform the rotations in the given order.

Grade:

Developing Spatial Thinking

Name:

Class: Section: Date:

Rotate the objects shown below by the indicated amount. Sketch the result in the space provided. Make sure you perforn the rotations in the given order.

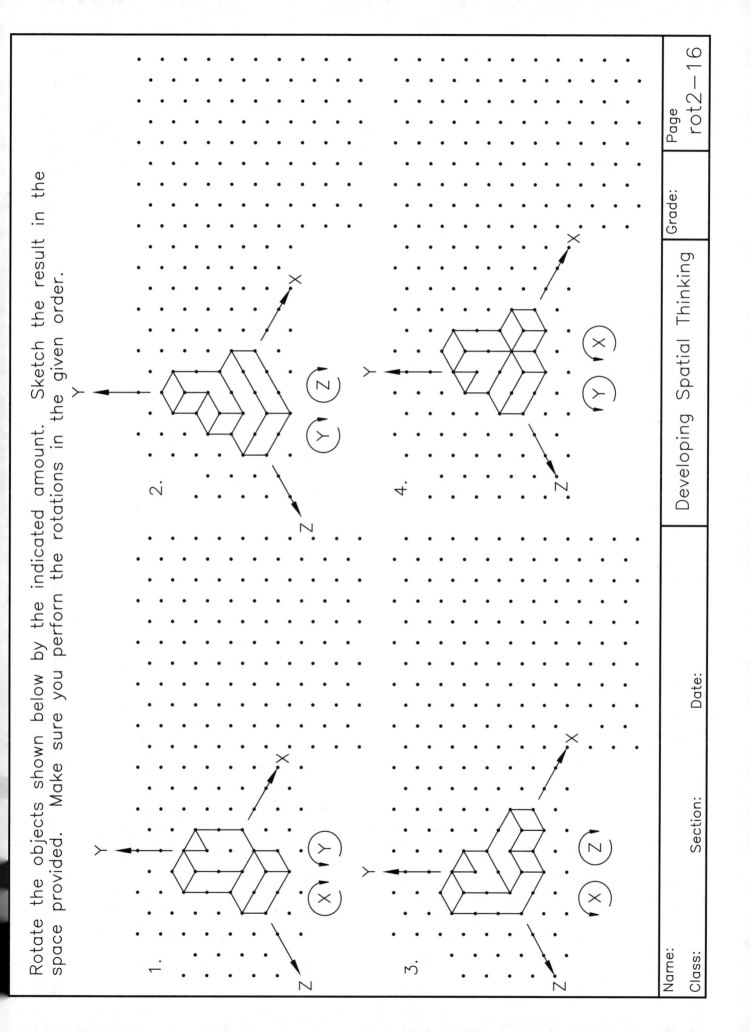

1.

2.

3.

4.

Name:

Class: Section: Date:

Developing Spatial Thinking

Grade:

Page
rot2—16

Rotate the objects shown below by the indicated amount. Sketch the result in the space provided. Make sure you perform the rotations in the given order.

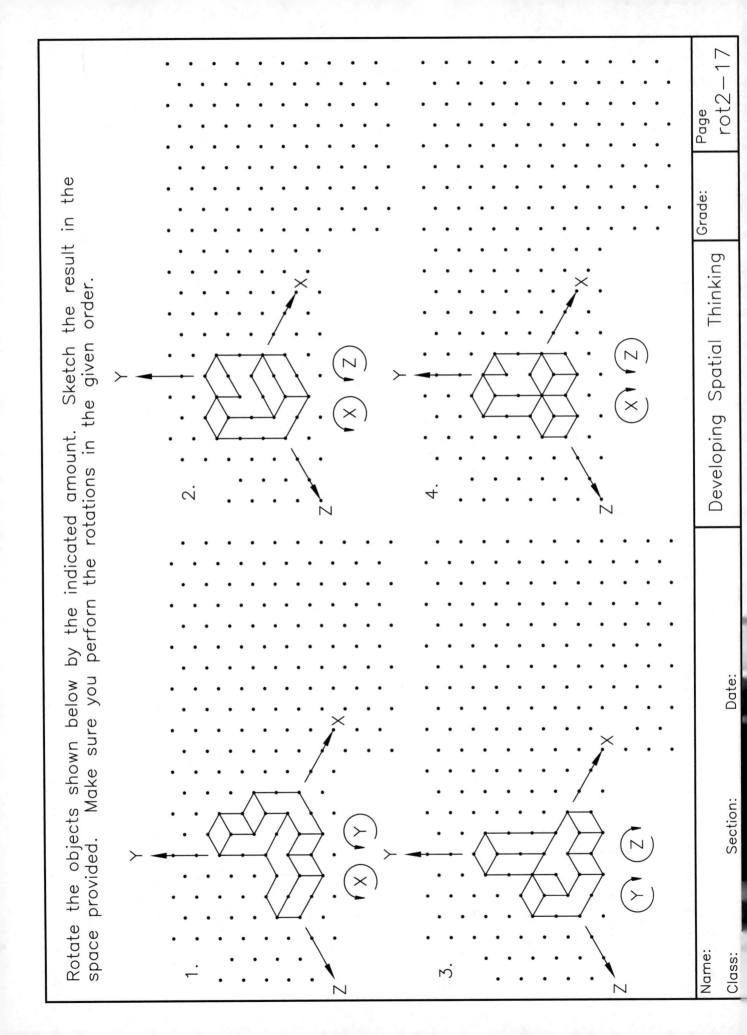

1.

2.

3.

4.

Grade:

Developing Spatial Thinking

Name:

Class:

Section:

Date:

Object Reflections and Symmetry

When an object is reflected across across a plane, the result is two separate objects that are "mirror images" of one another on both sides of the plane of reflection.

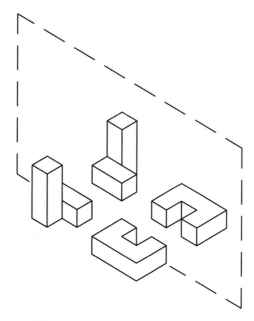

3-D Objects and Their Reflections Across the Indicated Plane

Each point on the reflected object corresponds to a point on the original object.

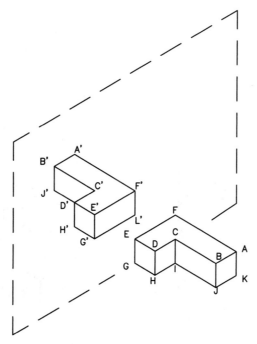

Object Points and Corresponding Points on Reflected Image

When drawing a reflection, imagine that the space on the opposite side of the plane of reflection is defined by the points from isometric dot paper. Points defining the object can be reflected through this plane, and corresponding edges and surfaces drawn. Points extend outward from the plane in both direction --- points that are "closer" to you on the original object will be "farther" from you on the reflected image and vice versa.

An object is said to be symmetrical if a plane can cut it so that the part of the object on one side of the plane is a mirror image of the part on the other side of a plane.

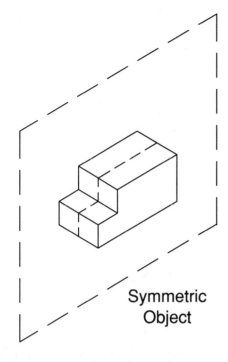

Symmetric
Object

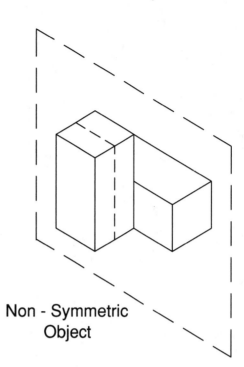

Non - Symmetric
Object

A plane of symmetry exists for a single object whereas a plane of reflection creates two separate objects.

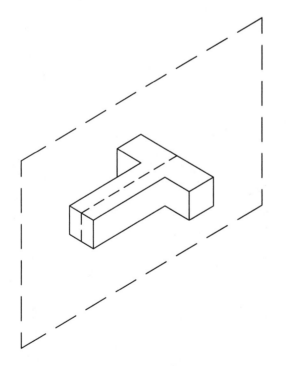

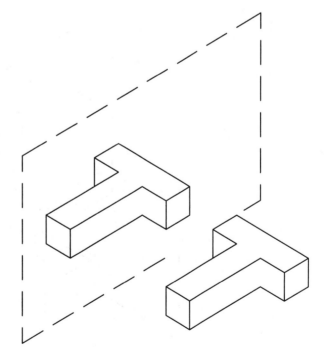

An object can have multiple planes of symmetry.

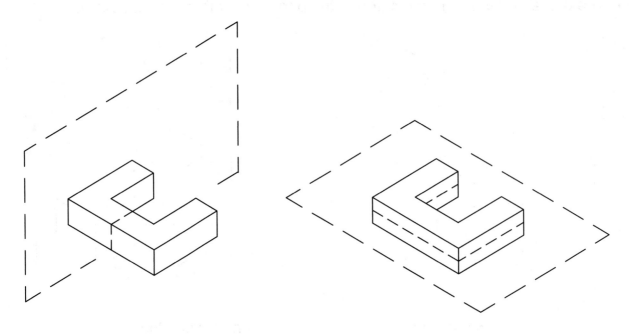

When an object is symmetric, a rotation of 180 degrees can achieve the same result as a reflection. In this case, the plane of symmetry and the plane of reflection must be perpendicular to one another. The axis of rotation is the line formed by the intersection of these two planes.

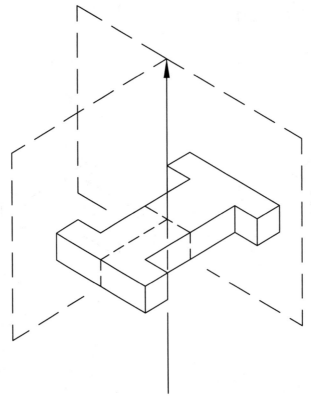

For the objects below, sketch the reflection across the indicated plane in the space provided. Sketch also the rotational axis about which the object could be rotated by 180° to achieve the same result.

1.

2.

3.

4.

Name:

Class:

Section:

Date:

Developing Spatial Thinking

Grade:

Page: reflx/sym−01

For the objects below, sketch the reflection across the indicated plane in the space provided. Sketch also the rotational axis about which the object could be rotated by 180° to achieve the same result.

1.

2.

3.

4.

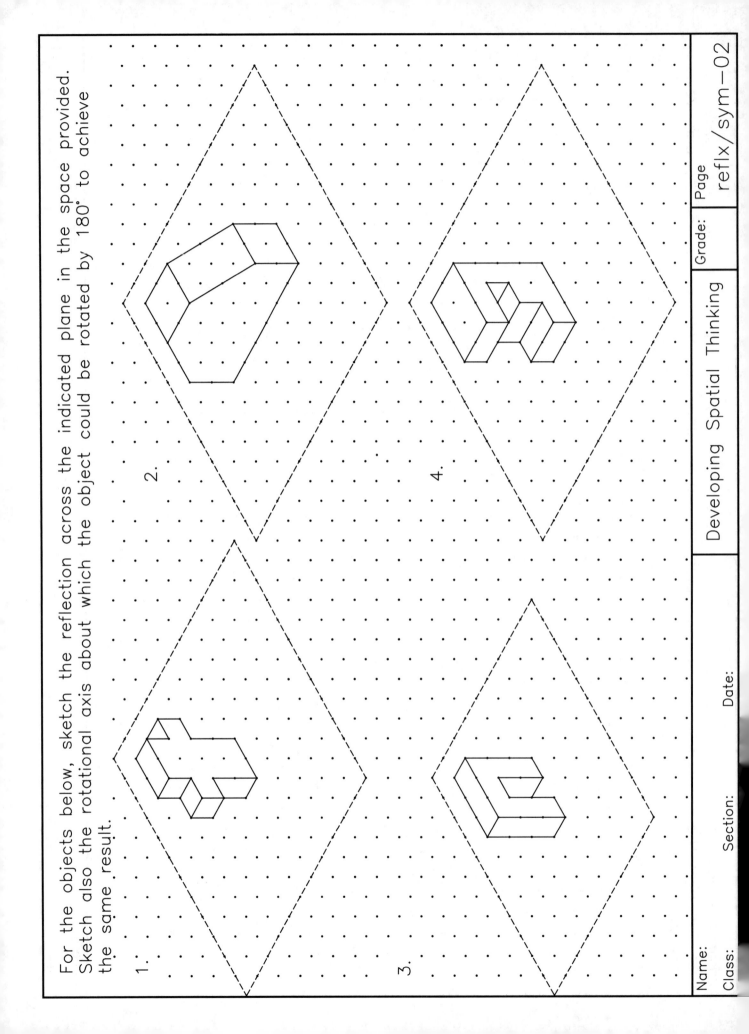

Name:

Class:

Section:

Date:

Developing Spatial Thinking

Grade:

Page
reflx/sym-02

For the objects below, sketch the reflection across the indicated plane in the space provided. Sketch also the rotational axis about which the object could be rotated by 180° to achieve the same result.

1.

2.

3.

4.

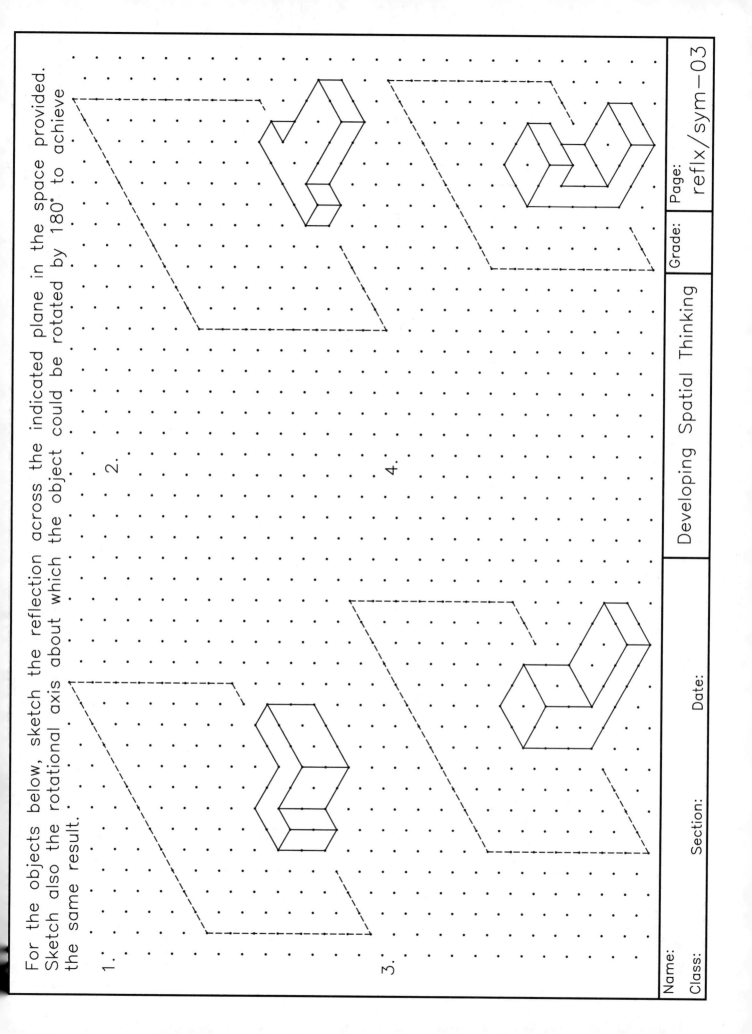

Name:

Class:

Section:

Date:

Developing Spatial Thinking

Grade:

Page:
reflx/sym−03

How many planes of symmetry do the objects shown below have?
Indicate your answer in the space provided. (Don't forget planes on the diagonal!)

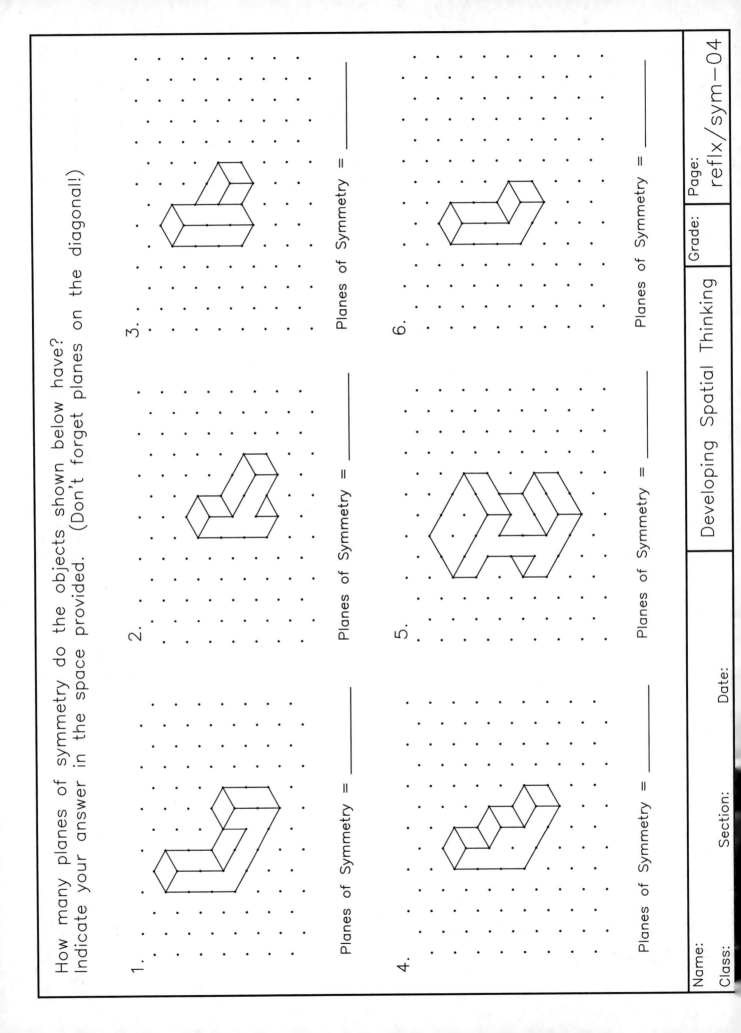

1.

Planes of Symmetry = _____

2.

Planes of Symmetry = _____

3.

Planes of Symmetry = _____

4.

Planes of Symmetry = _____

5.

Planes of Symmetry = _____

6.

Planes of Symmetry = _____

Developing Spatial Thinking

Page: reflx/sym-04

Grade:

Name:

Class:

Date:

Section:

How many planes of symmetry do the objects shown below have?
Indicate your answer in the space provided. (Don't forget planes on the diagonal!)

1.

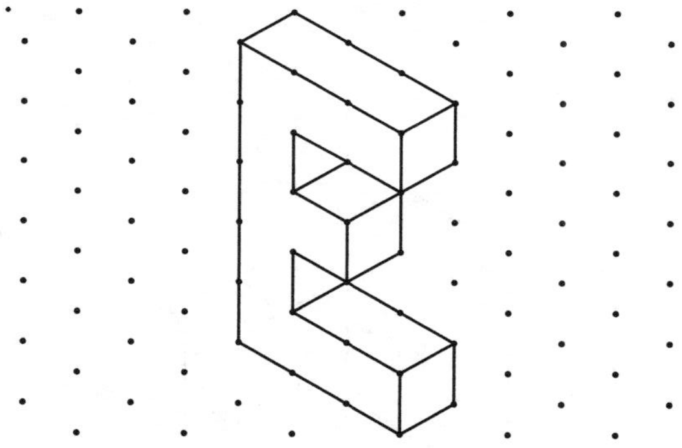

Planes of Symmetry = _____

2.

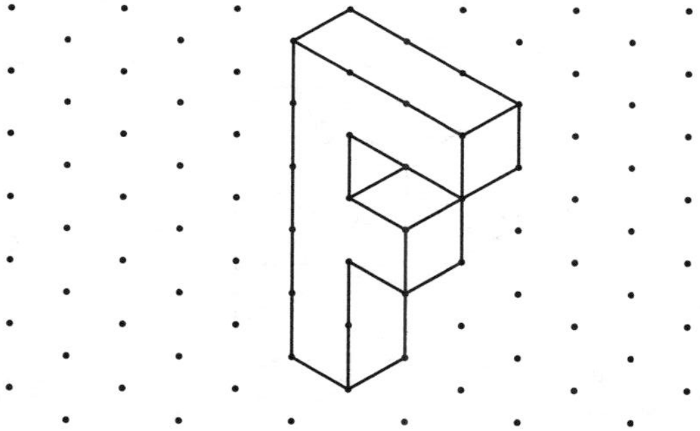

Planes of Symmetry = _____

3.

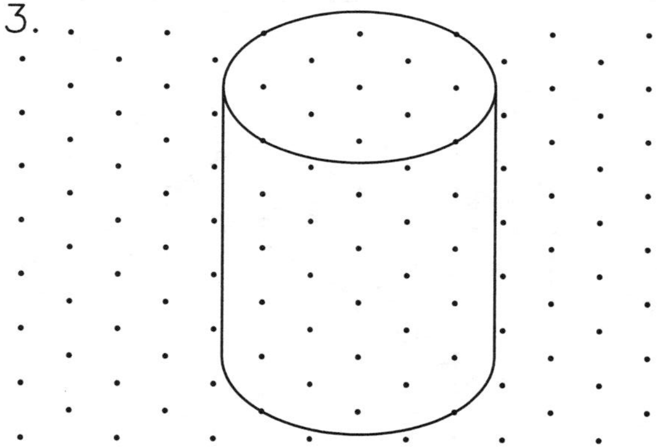

Planes of Symmetry = _____

4.

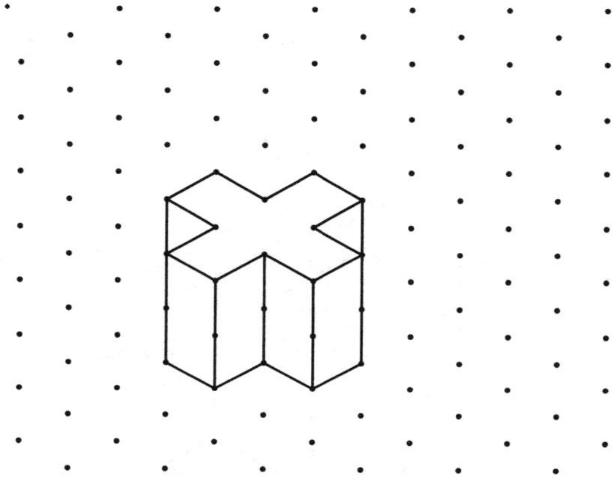

Planes of Symmetry = _____

5.

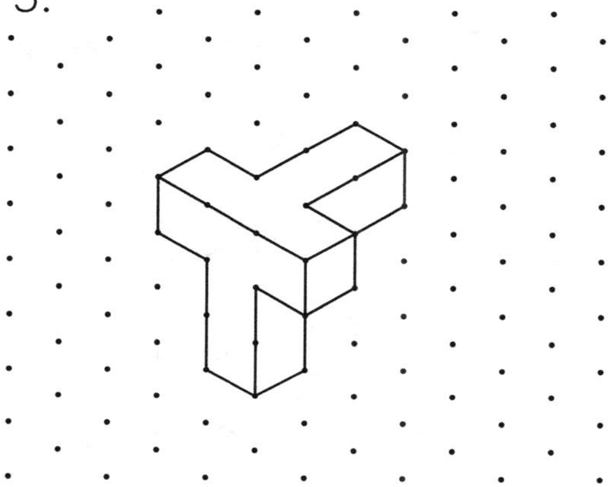

Planes of Symmetry = _____

6.

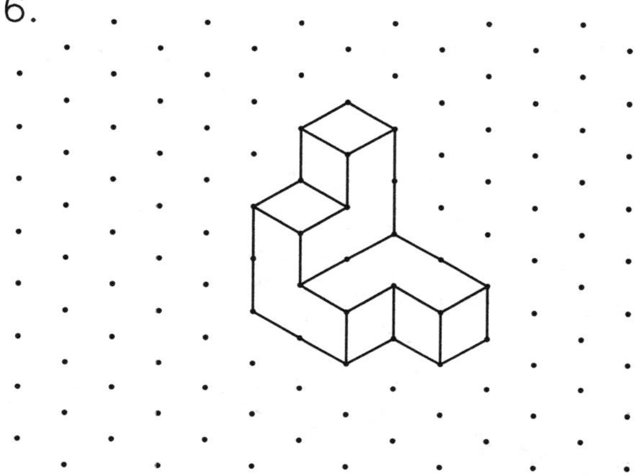

Planes of Symmetry = _____

| Name: | | Developing Spatial Thinking | Grade: | Page: |
| Class: | Date: | | Section: | reflx/sym-05 |

How many planes of symmetry do the objects shown below have?
Indicate your answer in the space provided. (Don't forget planes on the diagonal!)

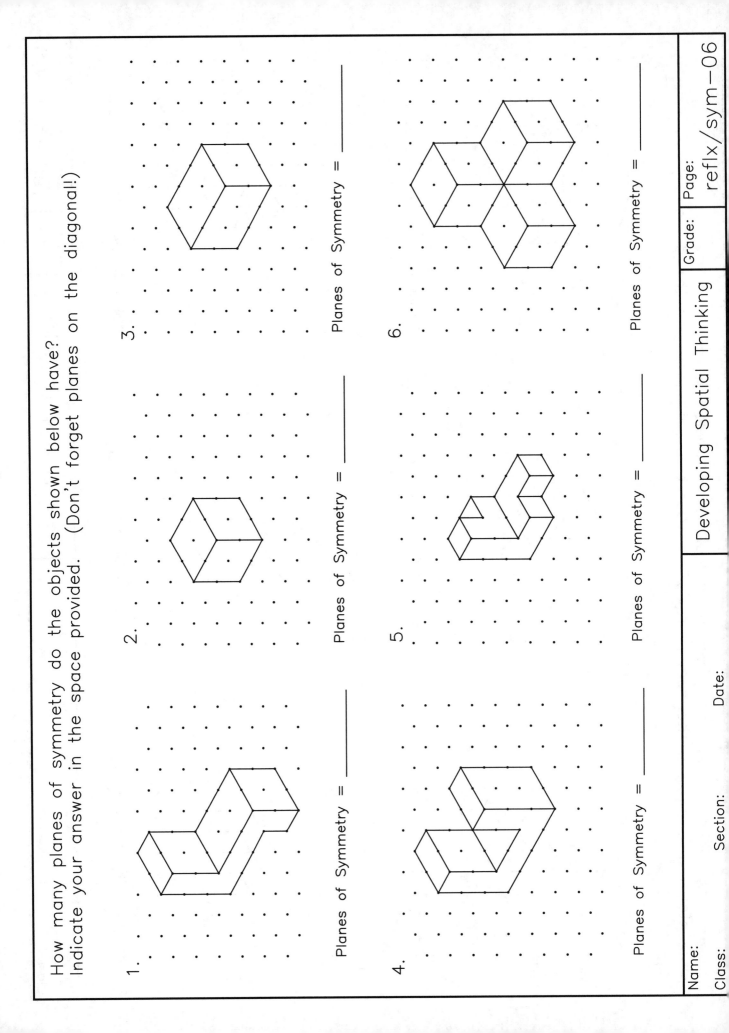

1.

Planes of Symmetry = _____

2.

Planes of Symmetry = _____

3.

Planes of Symmetry = _____

4.

Planes of Symmetry = _____

5.

Planes of Symmetry = _____

6.

Planes of Symmetry = _____

Sketch the reflected image of the objects below across the indicated plane. The sketch of the reflection is started for you.

1.

2.

3.

4.

Name: Class: Section: Date:	Developing Spatial Thinking	Grade:	Page: reflx/sym–07

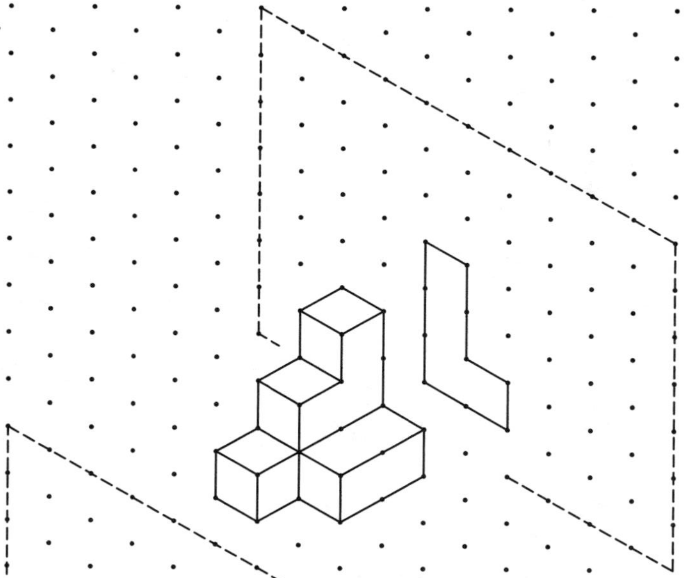

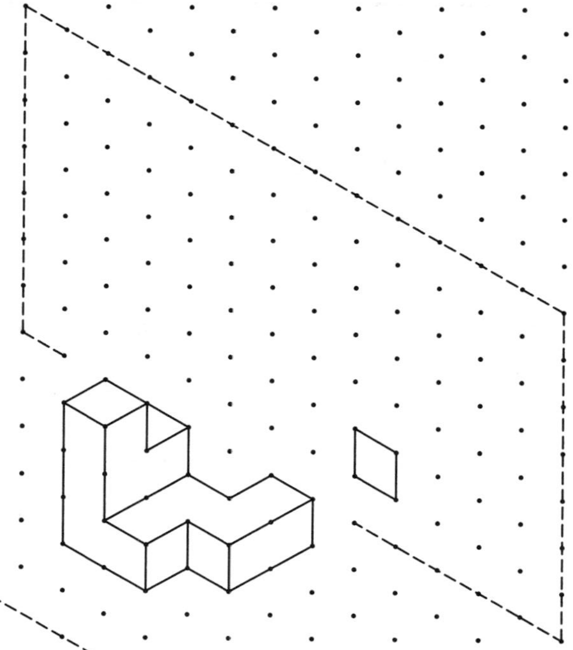

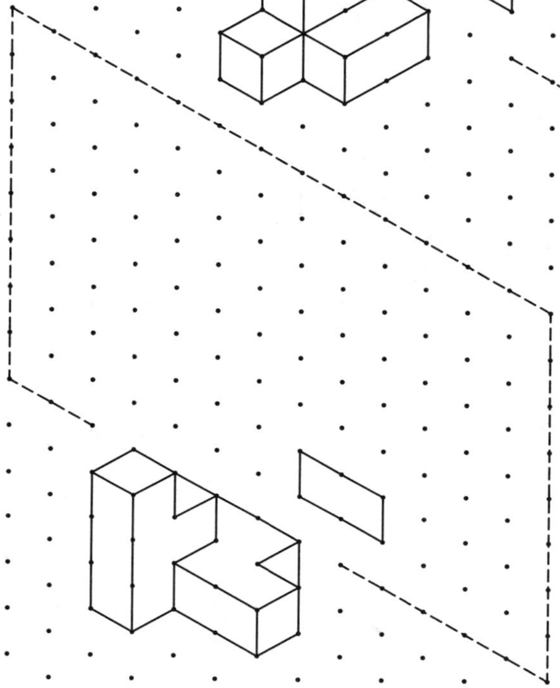

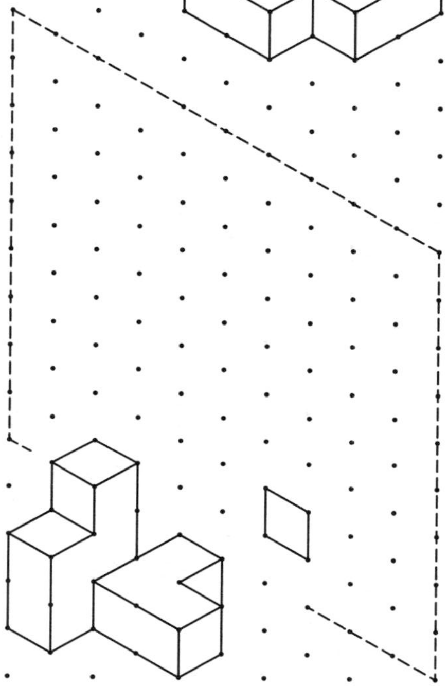

Sketch the reflected image of the objects below across the indicated plane. The sketch of the reflection is started for you.

1.

2.

3.

4.

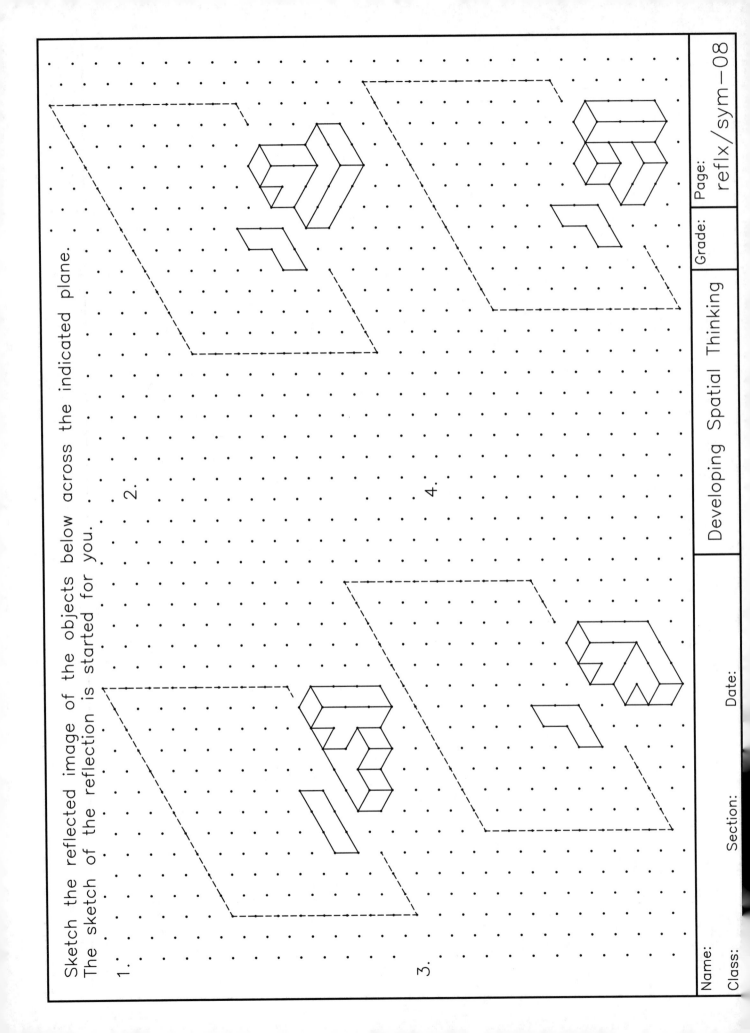

Sketch the reflected image of the objects below across the indicated plane.
The sketch of the reflection is started for you.

1.

2.

3.

4.

Name: Class: Section: Date:	Developing Spatial Thinking	Grade:	Page: reflx/sym–09

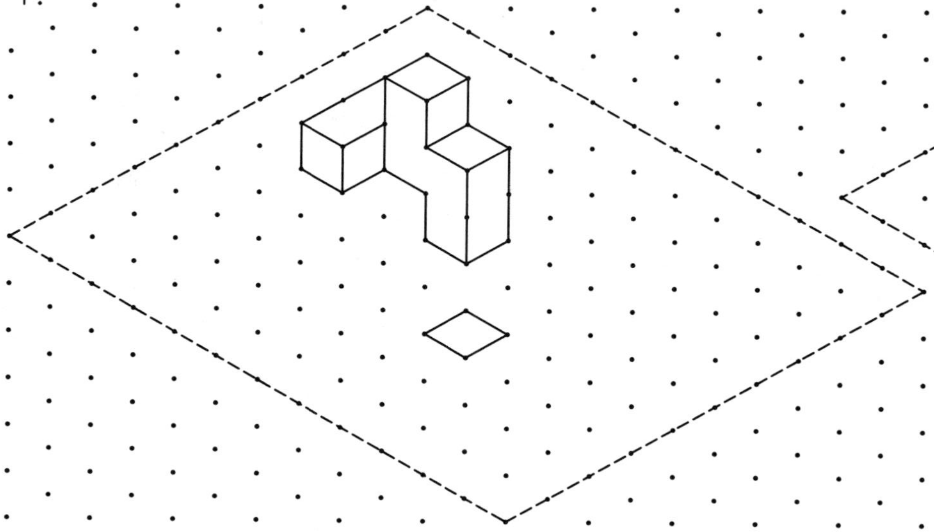

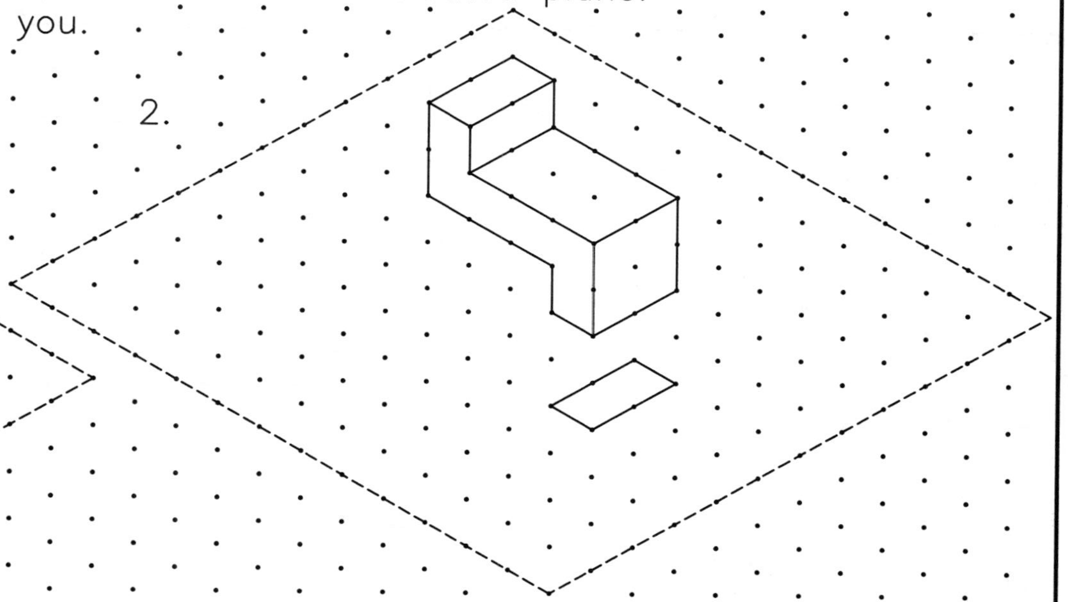

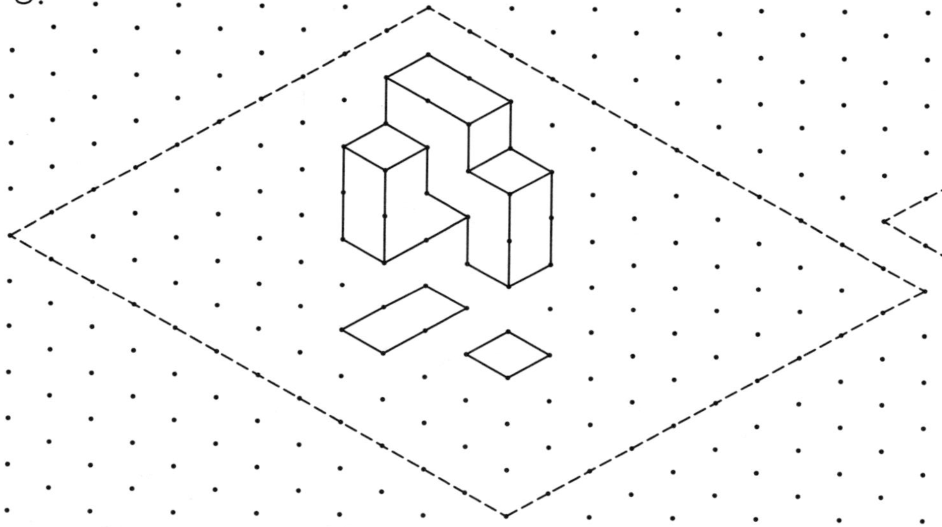

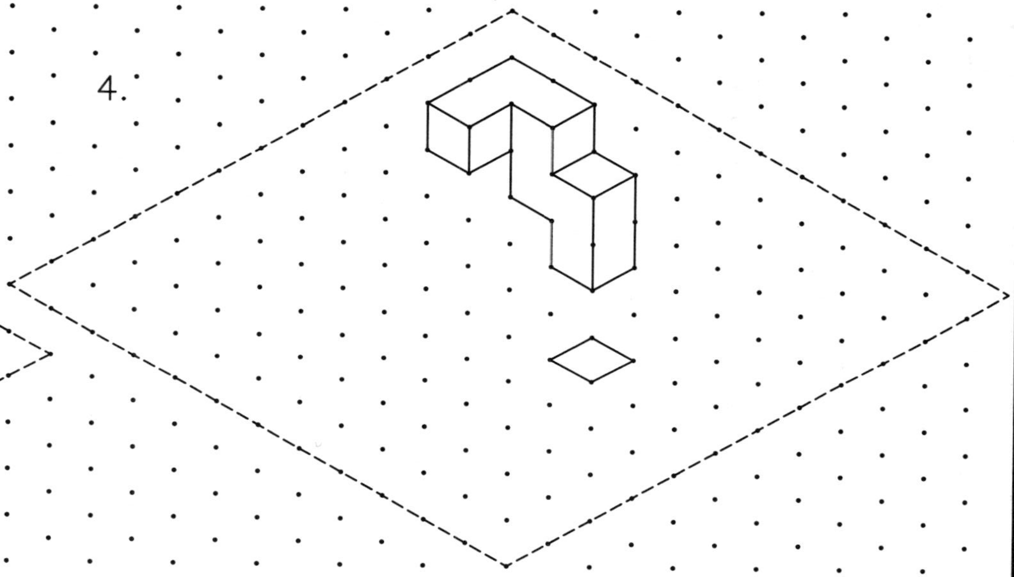

Sketch the reflected image of the objects below across the indicated plane.
The sketch of the reflection is started for you.

1.

2.

3.

4.

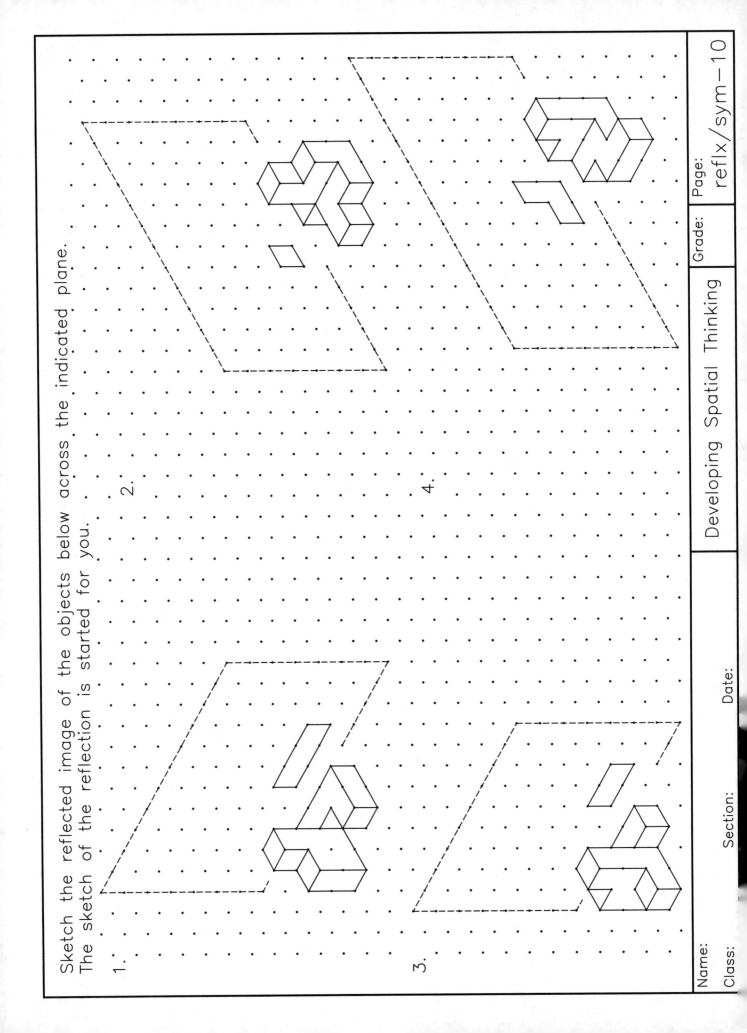

Developing Spatial Thinking

Page: reflx/sym-10

Grade:

Name:

Class:

Section:

Date:

An object and a plane of reflection are shown on the left below.
Select the reflected image from the choices given.

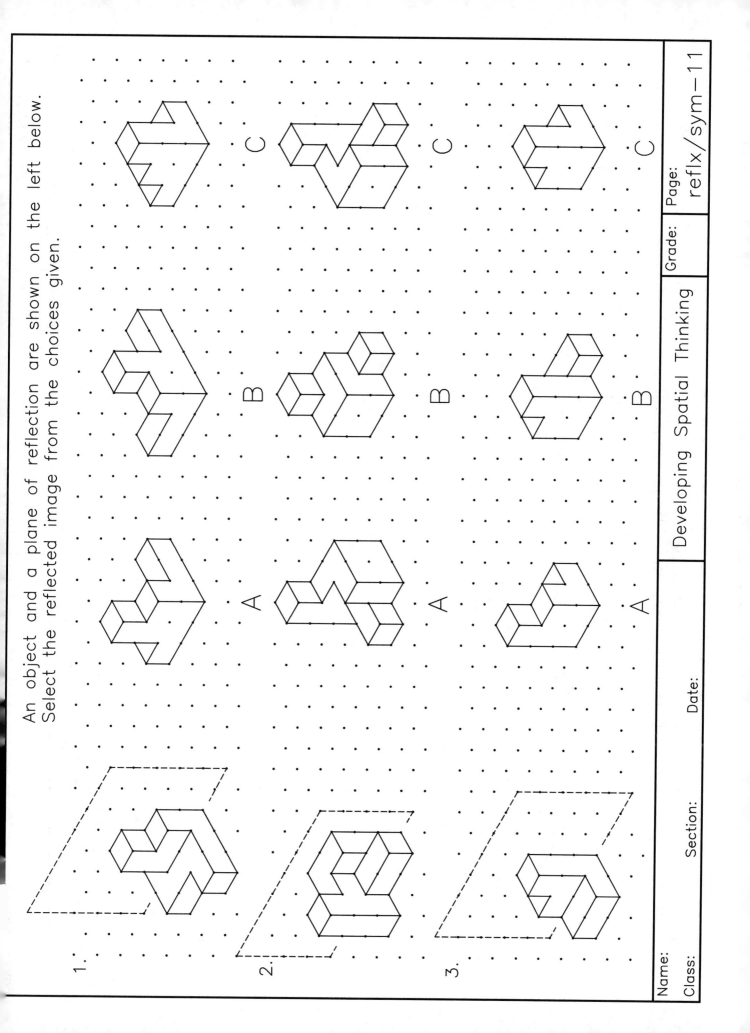

1.

2.

3.

A

A

A

B

B

B

C

C

C

An object and a plane of reflection are shown on the left below.
Select the reflected image from the choices given.

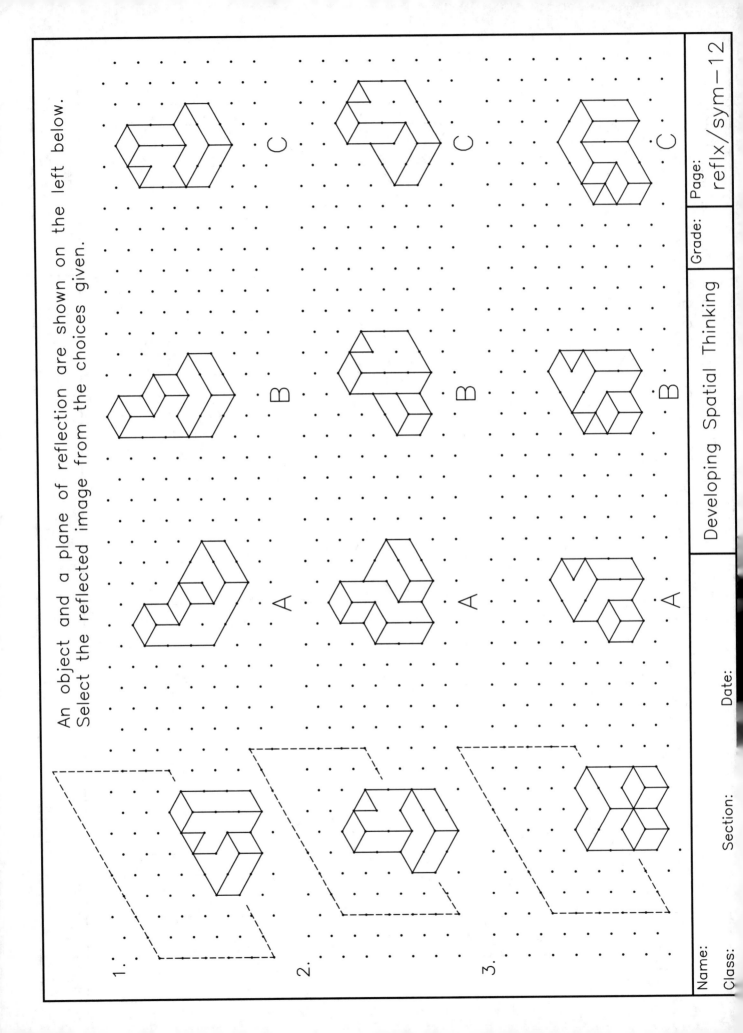

1.

2.

3.

A

B

C

Grade:

Developing Spatial Thinking

Name:

Class:

Section:

Date:

An object and a plane of reflection are shown on the left below.
Select the reflected image from the choices given.

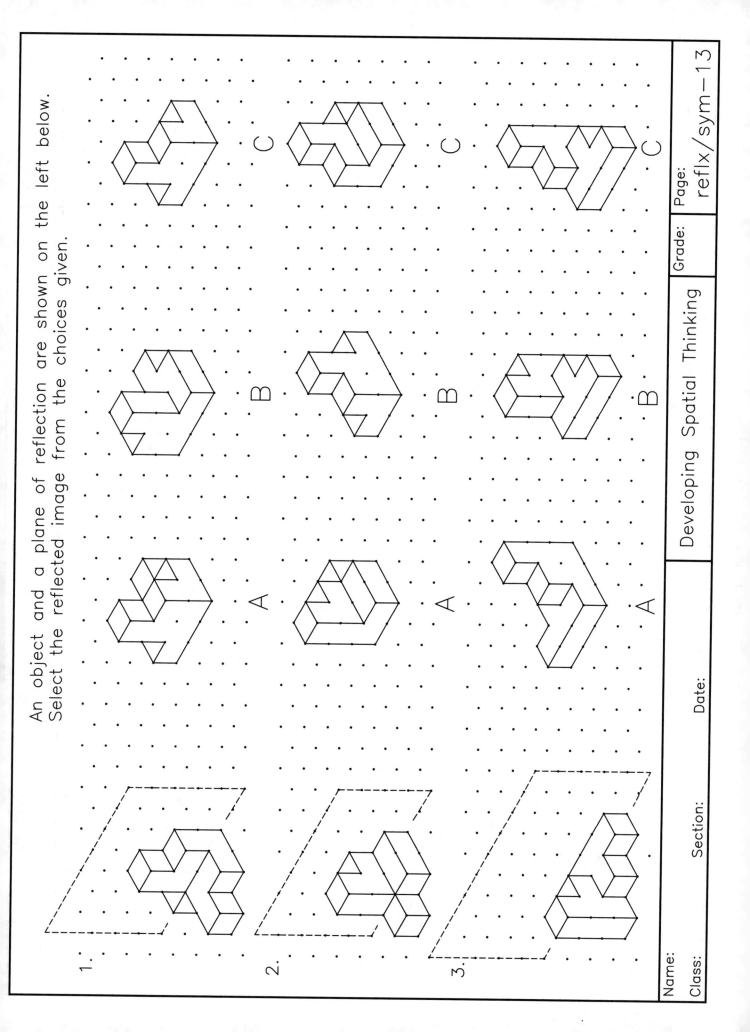

1.

2.

3.

A

A

A

B

B

B

C

C

C

An object and a plane of reflection are shown on the left below.
Select the reflected image from the choices given.

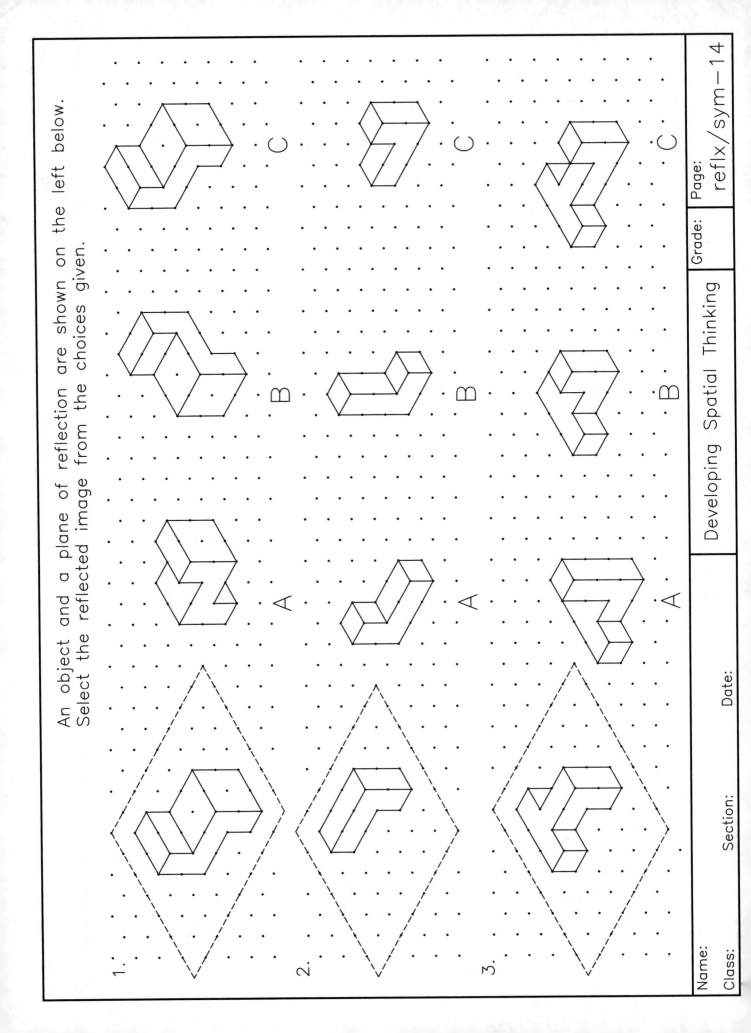

1.

2.

3.

A

B

C

A

B

C

A

B

C

Developing Spatial Thinking

Name:

Class:

Section:

Date:

Grade:

Page:

reflx/sym-14

An object and a plane of reflection are shown on the left below.
Select the reflected image from the choices given.

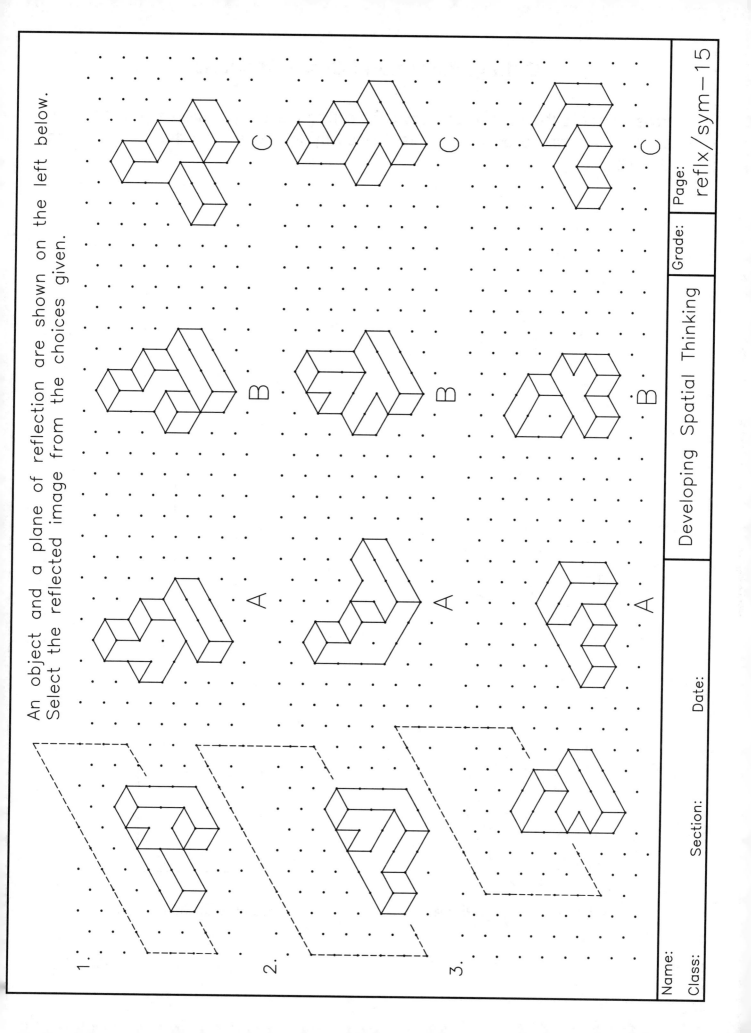

1.
2.
3.

A   B   C

Name:
Class: Section: Date:

Cutting Planes and Cross Sections

A cross section is the intersection between a cutting plane and a solid object. The result is a 2-D shape whose boundaries are defined by the edges and the surfaces of the original object.

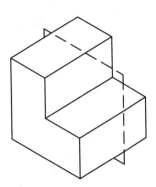
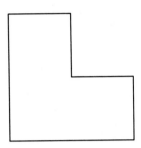

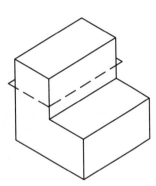

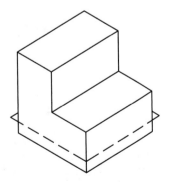

Object and Cutting Plane   Resulting Cross Section

When a plane cuts an object, the shape of the resulting cross section depends on the orientation of the cutting plane and the object with respect to one another.

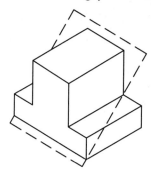
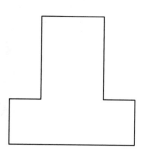

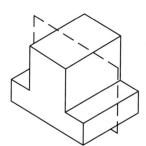
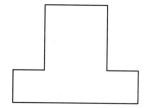

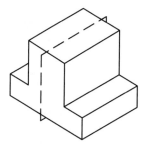

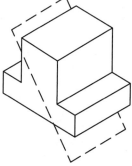

Object and Cutting Plane

Resulting Cross Section

Objects can produce several cross sections.

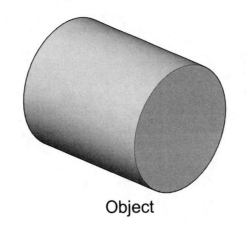

Object

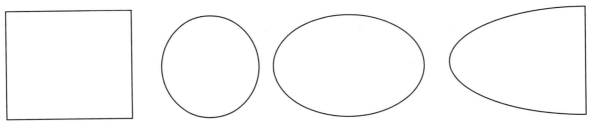

Possible Cross Sections

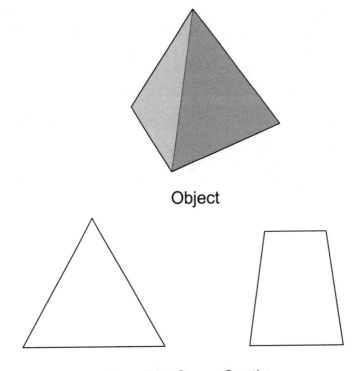

Object

Possible Cross Sections

To visualize the cross section that a cutting plane produces with an object, imagine the cutting plane extending through the object. As the plane "cuts" through a surface, it will intersect with it and a "boundary" of the cross section will be created. The boundary edges must be on the surfaces of the object itself and parallel to the edges of the cutting plane. After you have constructed the boundaries defining the cross section, mentally rotate the result so that it lies in the viewing plane.

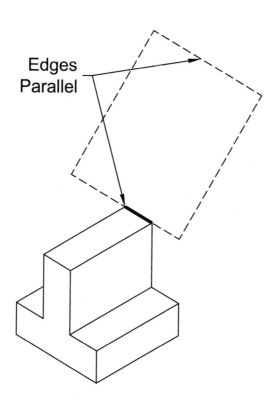

Edges Parallel

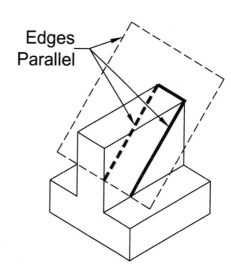

Edges Parallel

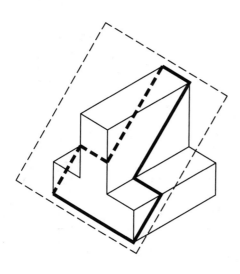

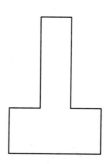

Cross Section Rotated to Viewing Plane

For the objects shown on the left below with an indicated cutting plane. Circle the letter corresponding to the correct cross-section that is obtained.

1.

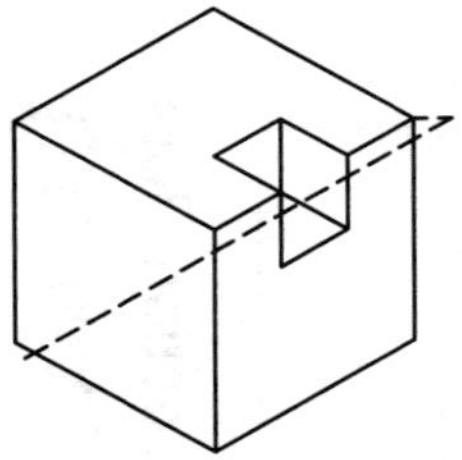

A 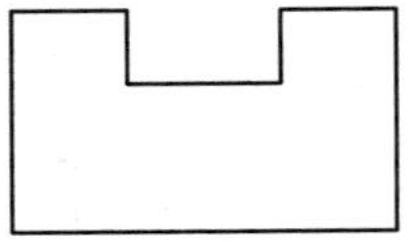

B 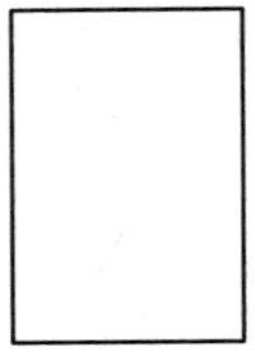

C 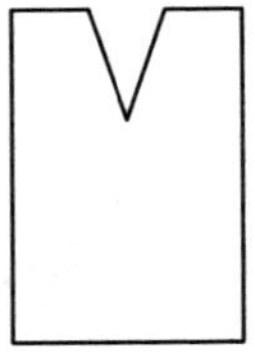

D 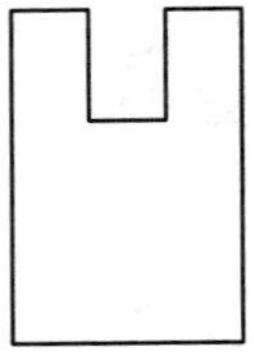

2.

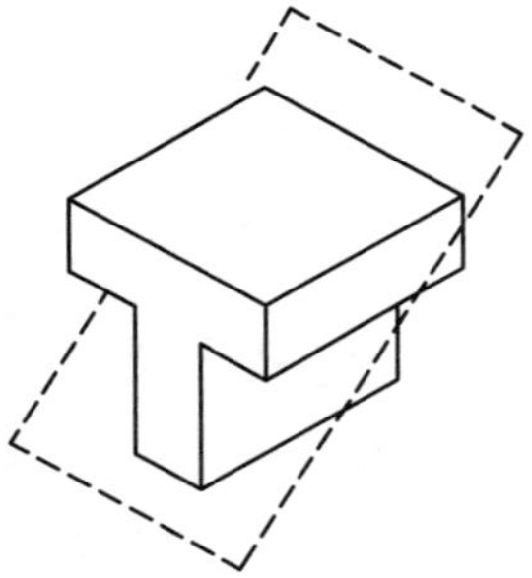

A 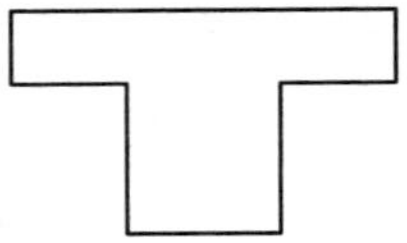

B 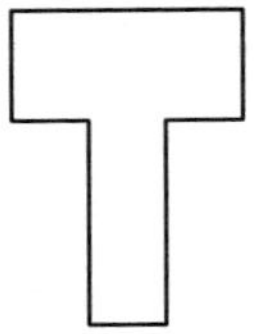

C 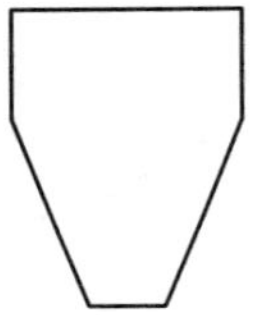

D 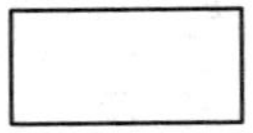

3.

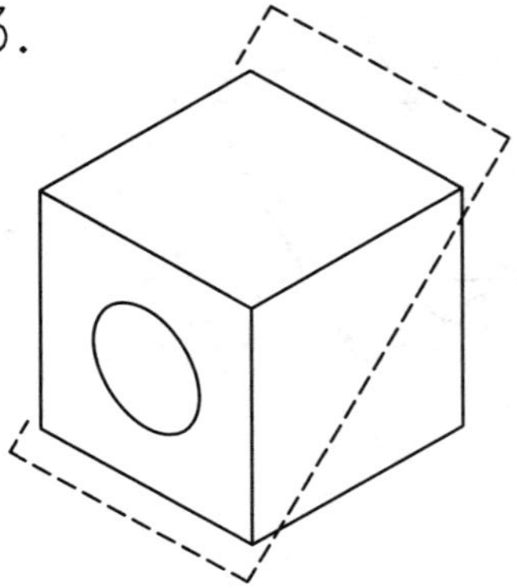

A 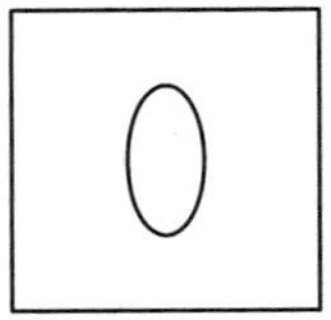

B 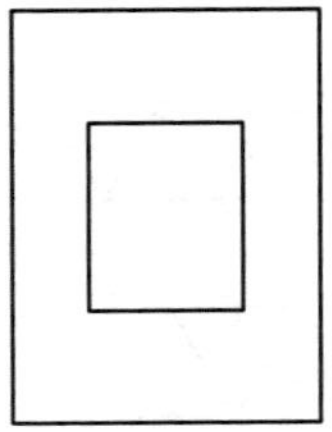

C 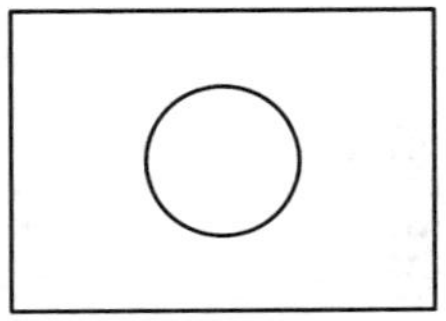

D 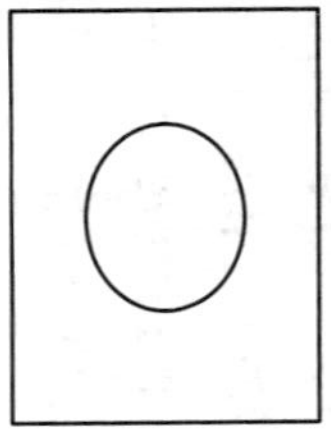

Name: Class: Section: Date:	Developing Spatial Thinking	Grade:	Page cp/cs−01

For the objects shown on the left below with an indicated cutting plane. Circle the letter corresponding to the correct cross-section that is obtained.

1.

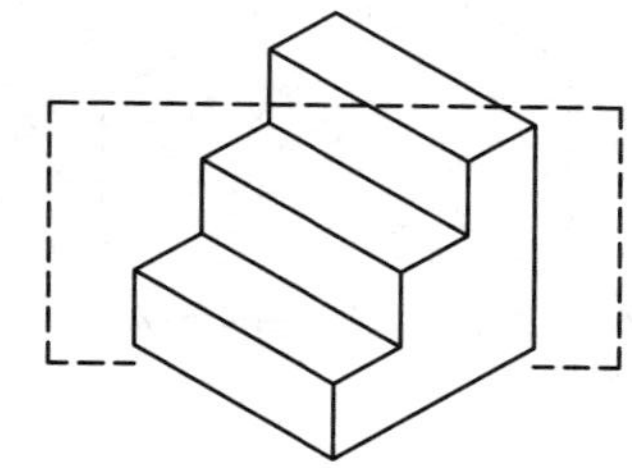

A 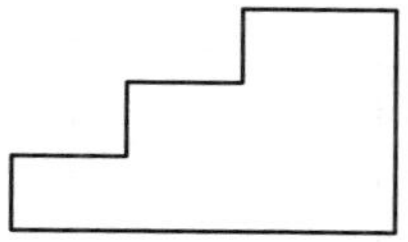   B 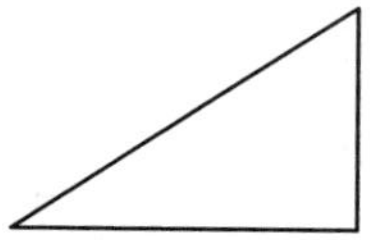   C 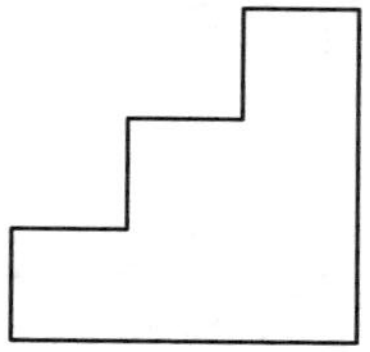   D

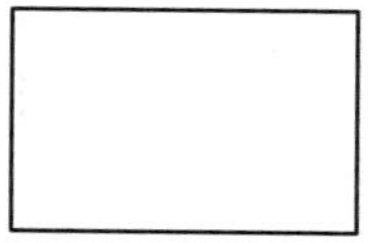

2.

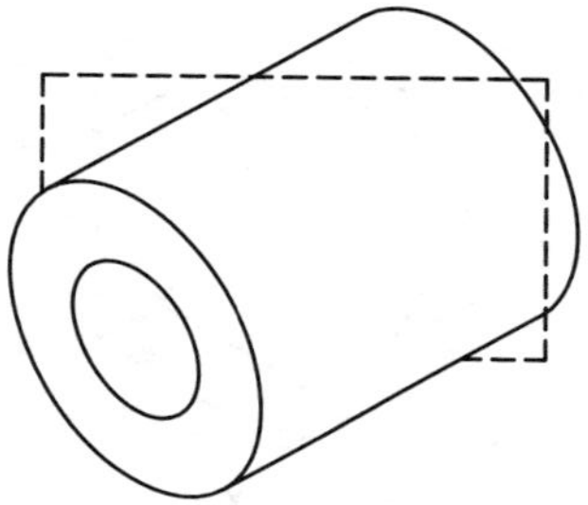

A 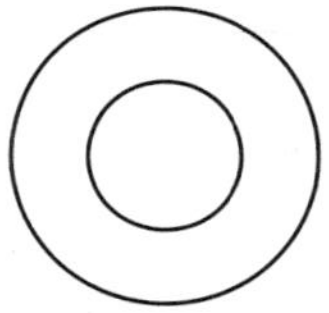   B 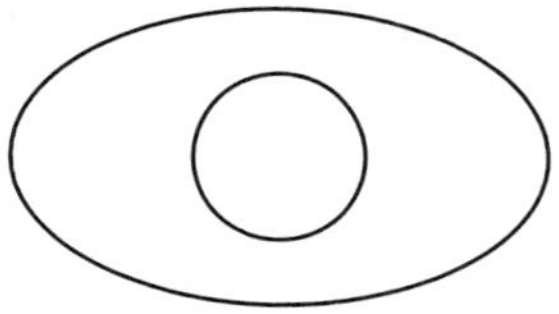   C 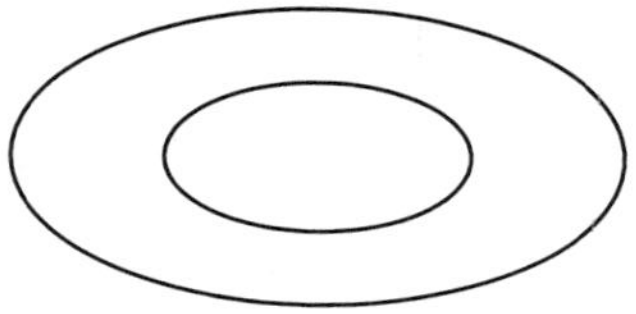   D

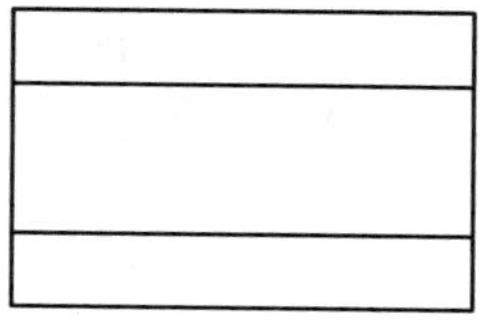

3.

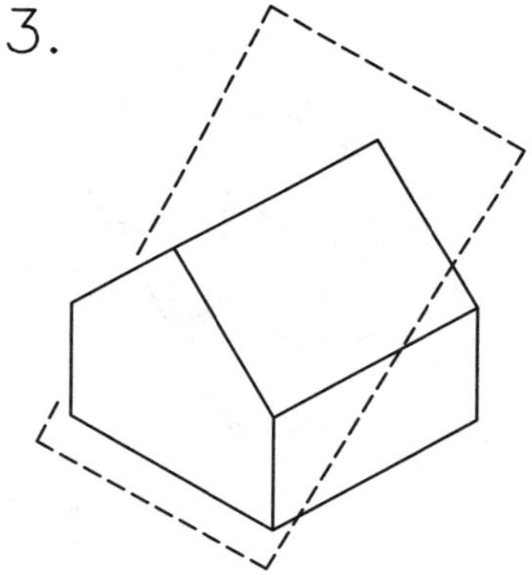

A 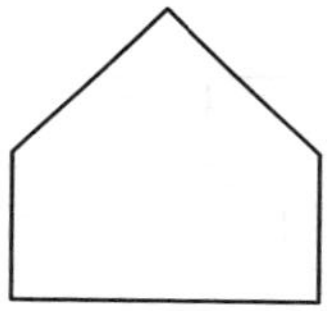   B 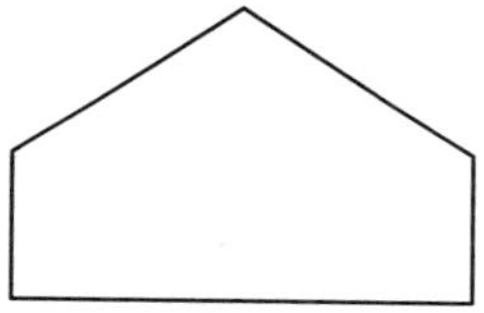   C 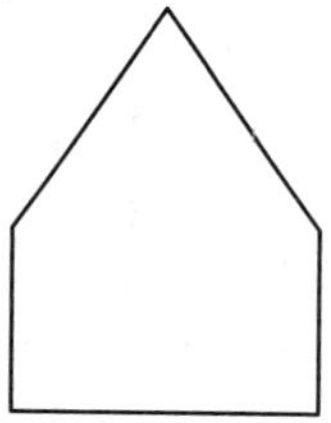   D

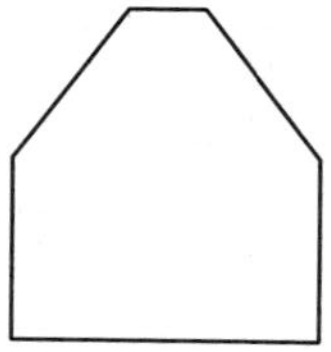

Name: Section: Date:

Class: Developing Spatial Thinking   Grade: Page
                                                                                 cp/cs—02

For the objects shown on the left below with an indicated cutting plane.
Circle the letter corresponding to the correct cross-section that is obtained.

1.

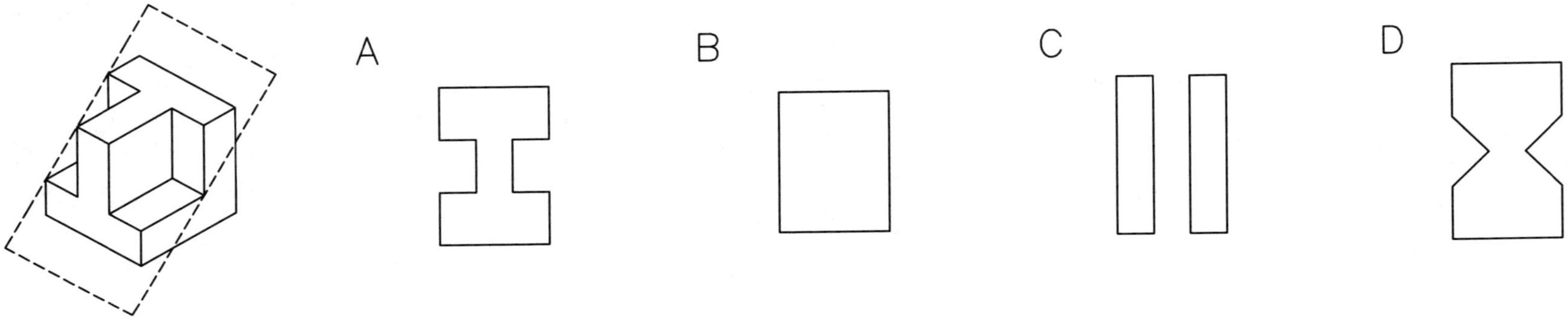

A   B   C   D

2.

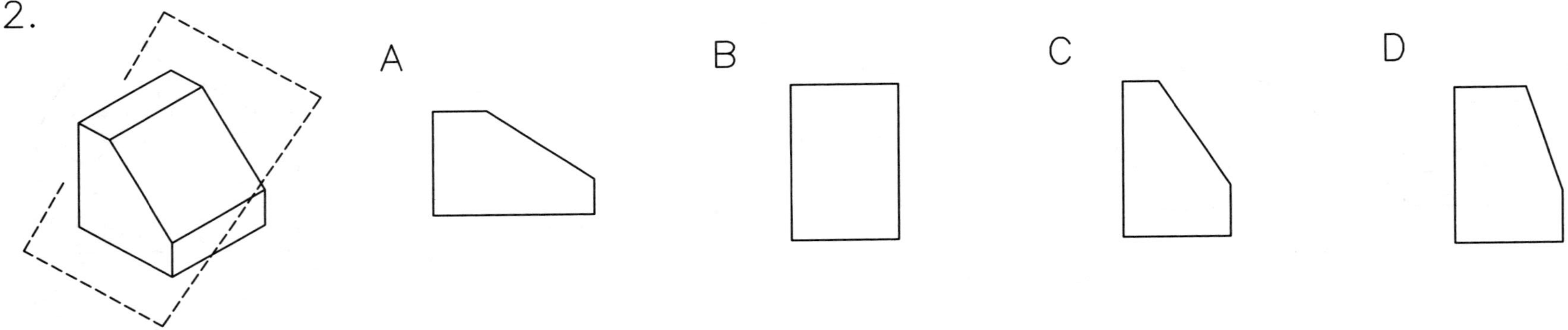

A   B   C   D

3.

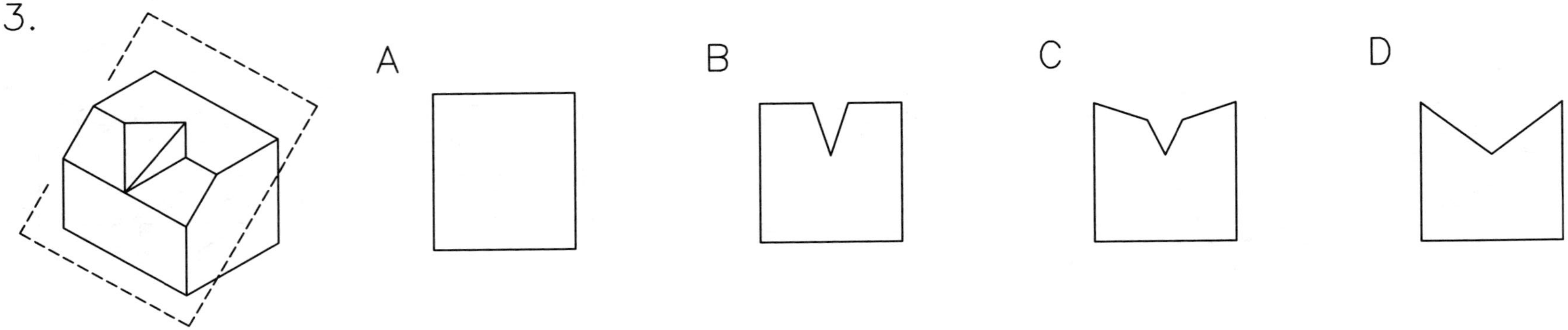

A   B   C   D

Grade:

Developing Spatial Thinking

Name: Date:

Class: Section:

For the objects shown on the left below with an indicated cutting plane. Circle the letter corresponding to the correct cross-section that is obtained.

1.

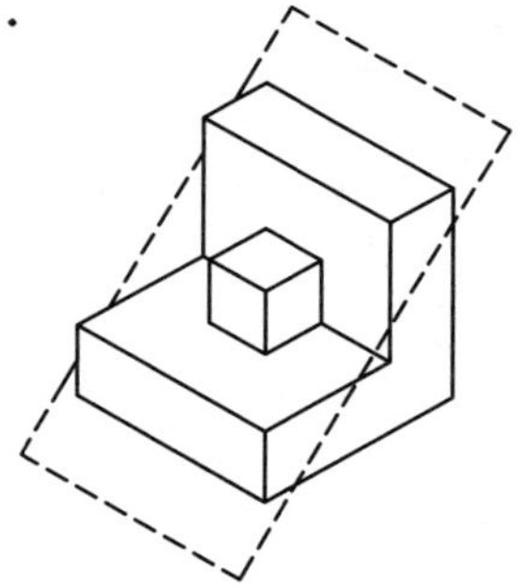

A 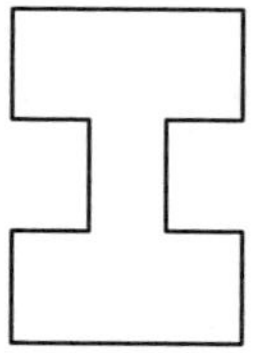   B 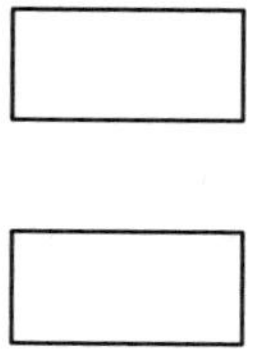   C   D

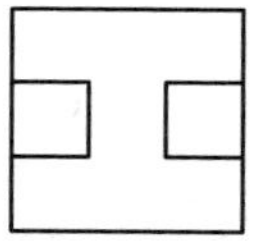

2.

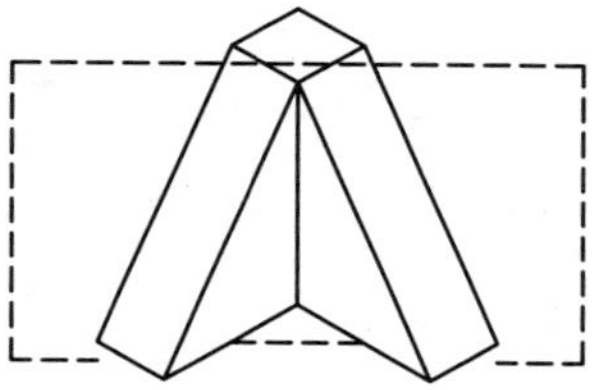

A 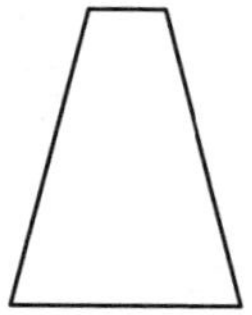   B 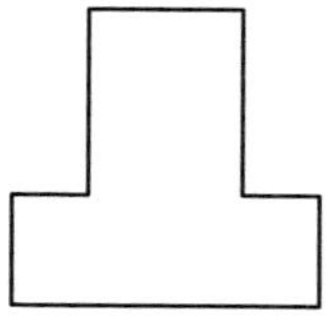   C 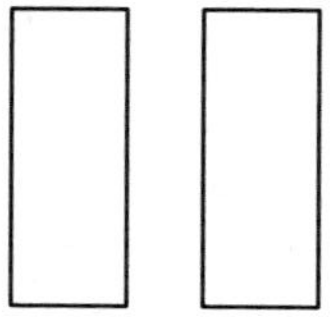   D

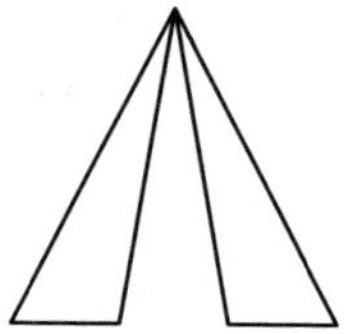

3.

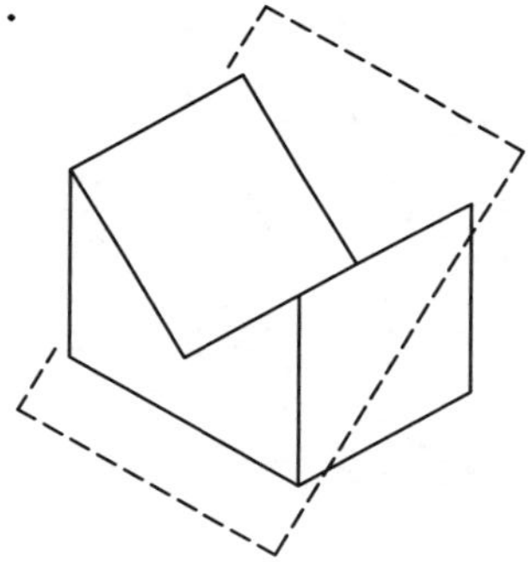

A 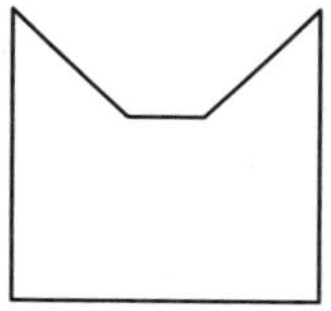   B 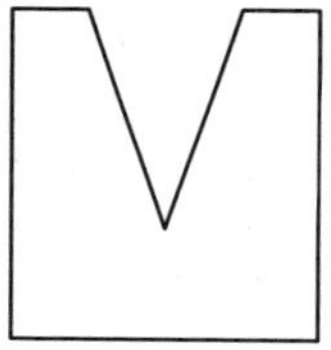   C 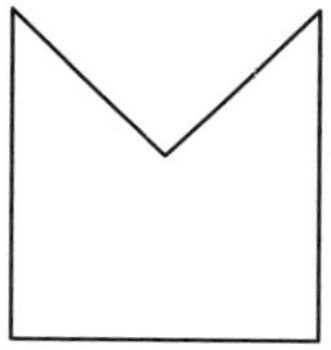   D

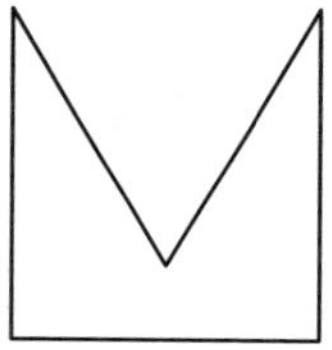

Match the cross-sections with the appropriate object and cutting plane for the problems shown below.

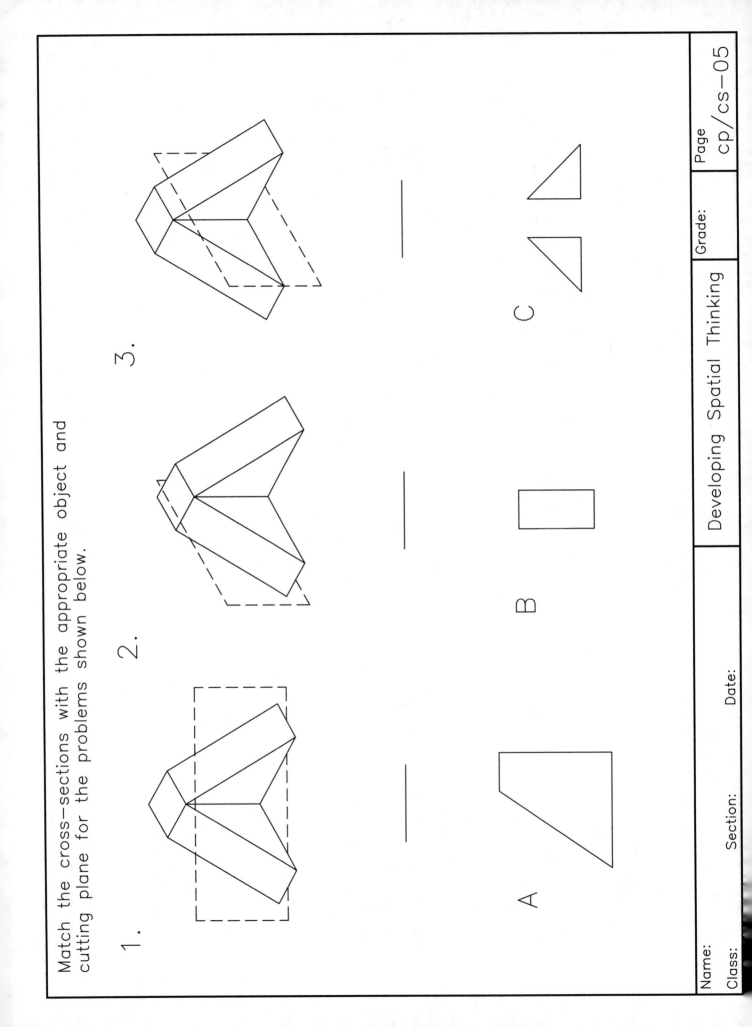

1.

2.

3.

A

B

C

Name: Section: Date:

Class:

Developing Spatial Thinking   Grade: Page
                                                 cp/cs—05

Match the cross-sections with the appropriate object and cutting plane for the problems shown below.

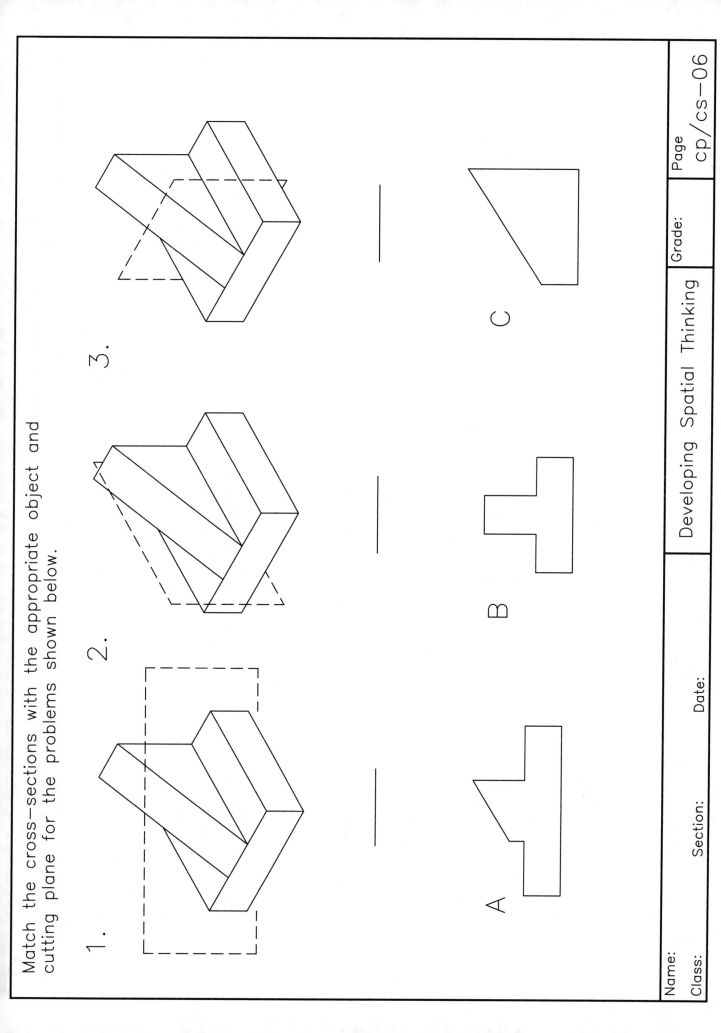

1.

2.

3.

A

B

C

Name:

Class:

Section:

Date:

Developing Spatial Thinking

Grade:

Page
cp/cs-06

Match the cross-sections with the appropriate object and
cutting plane for the problems shown below.

1.

2.

3.

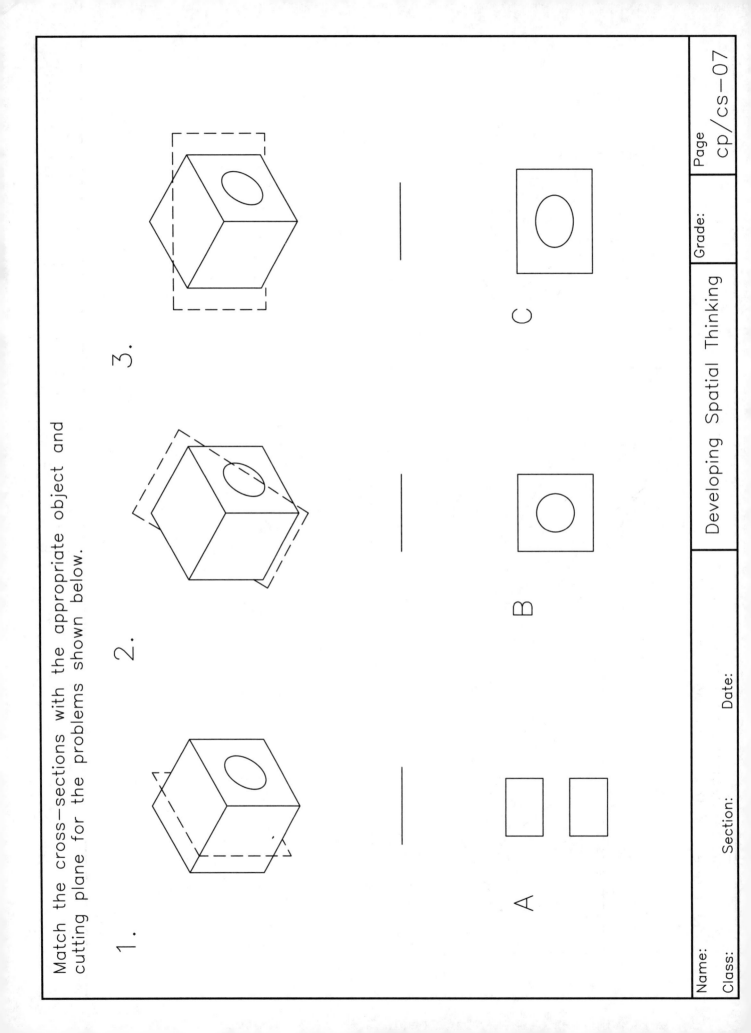

A

B

C

Grade:

Developing Spatial Thinking

Name:

Class:

Section:

Date:

Match the cross-sections with the appropriate object and cutting plane for the problems shown below.

1.

2.

3.

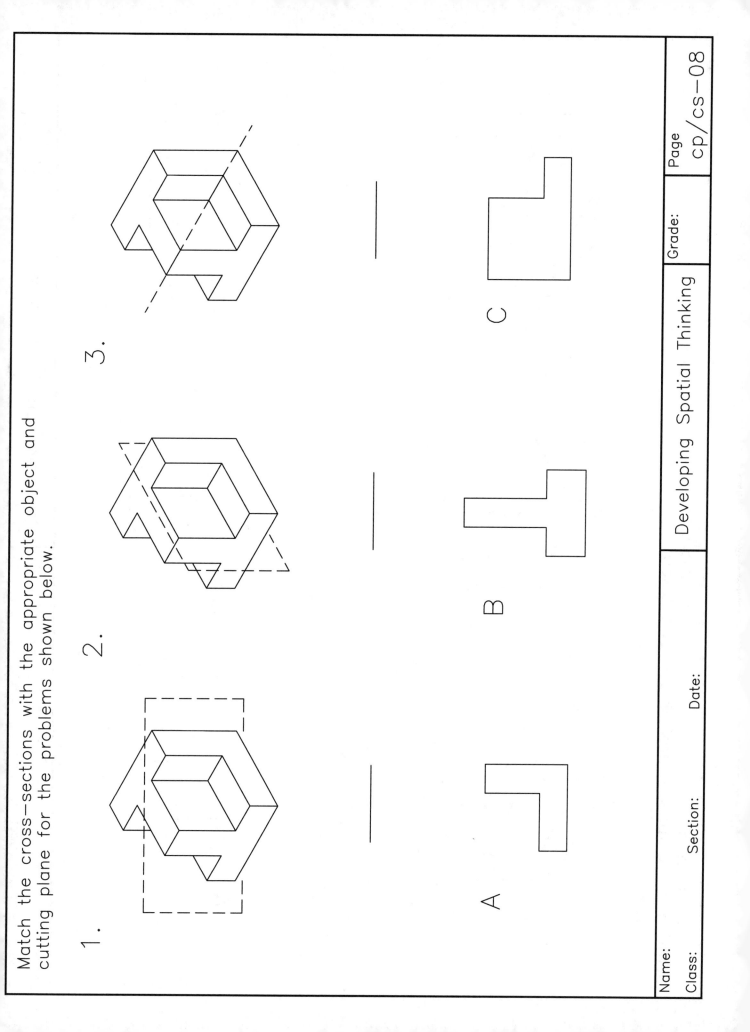

A

B

C

Developing Spatial Thinking

Page cp/cs-08

Name:

Class:

Section:

Date:

Grade:

Match the cross-sections with the appropriate object and cutting plane for the problems shown below.

1.

2.

3.

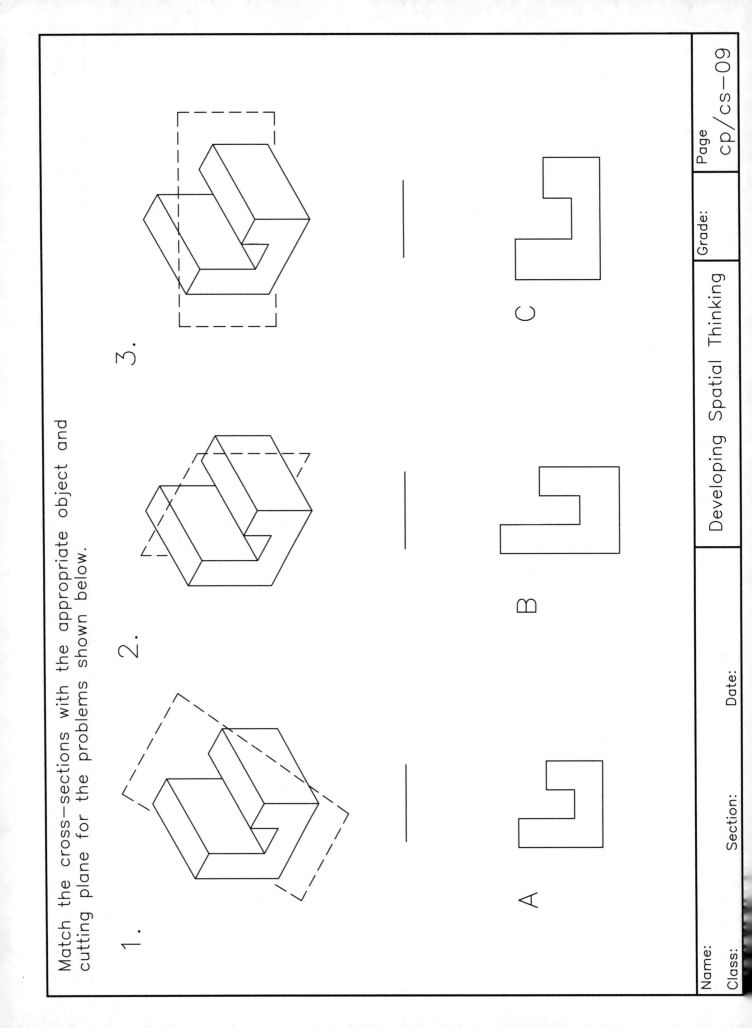

A

B

C

Grade:

Developing Spatial Thinking

Name:

Class: Section: Date:

Match the cross-sections with the appropriate object and cutting plane for the problems shown below.

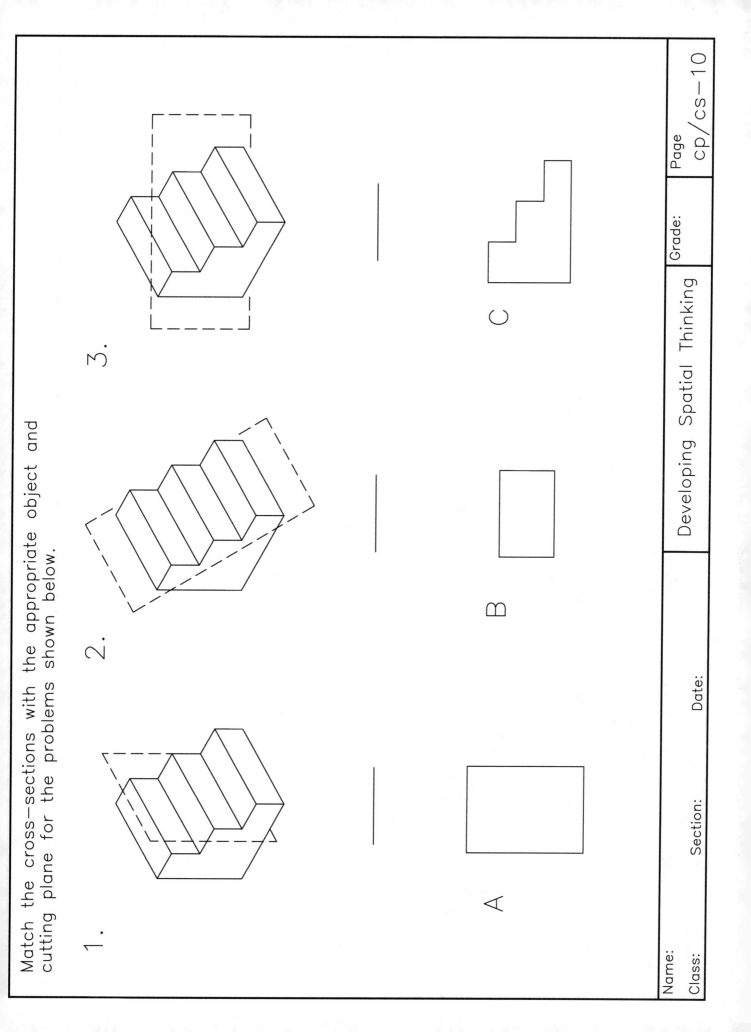

1.

2.

3.

A

B

C

Developing Spatial Thinking

Name:

Class:

Section:

Date:

Grade:

Page
cp/cs-10

Match the cross-sections with the appropriate object and cutting plane for the problems shown below.

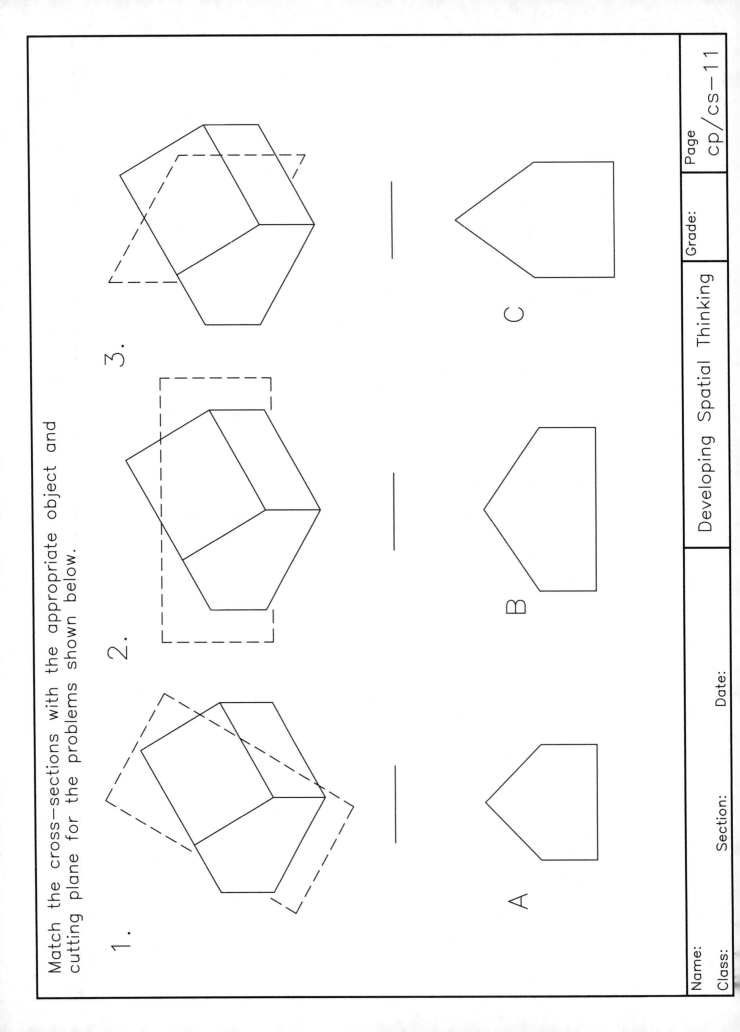

1.

2.

3.

A

B

C

Grade:

Developing Spatial Thinking

Name:

Class:

Section:

Date:

For the objects shown on the left below, circle the letters corresponding to all possible cross-sections that could be obtained by slicing it with an appropriate cutting plane. There could be more than one correct answer.

1.

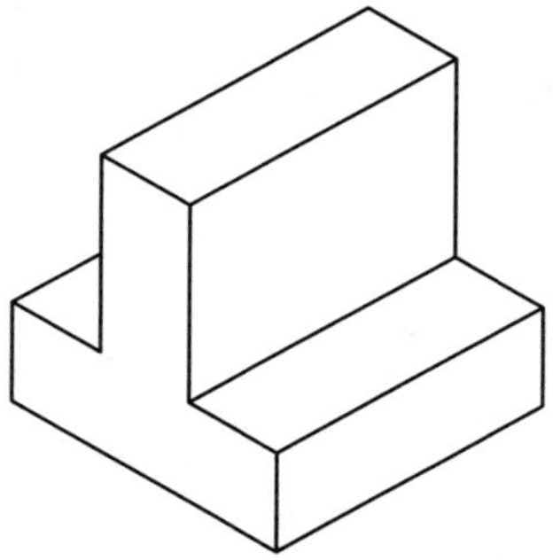

A 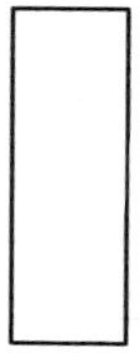   B 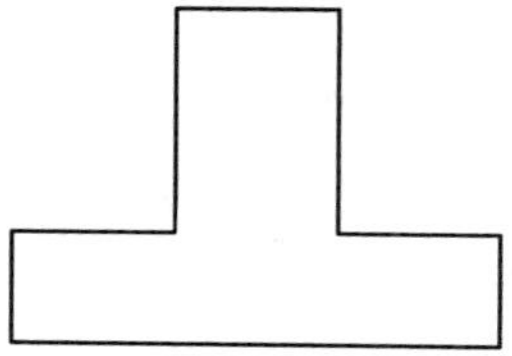   C 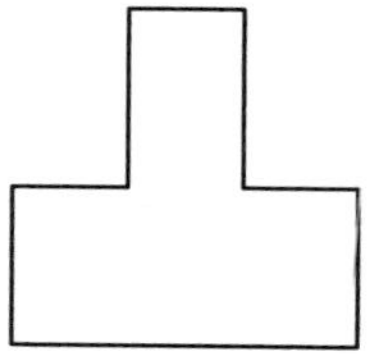   D

2.

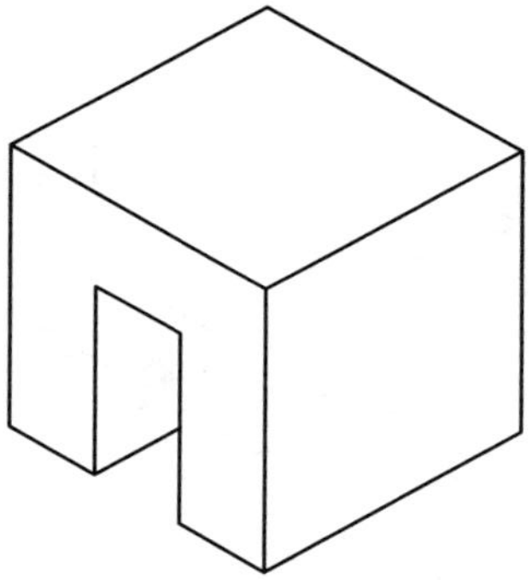

A 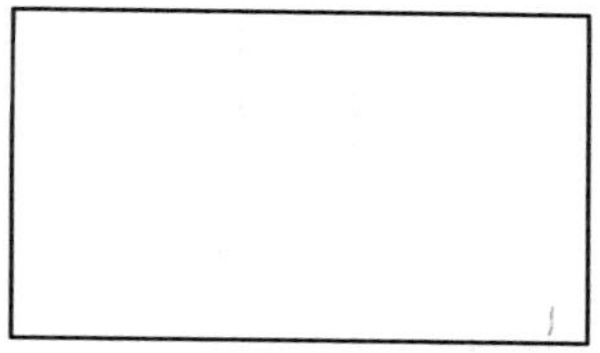   B 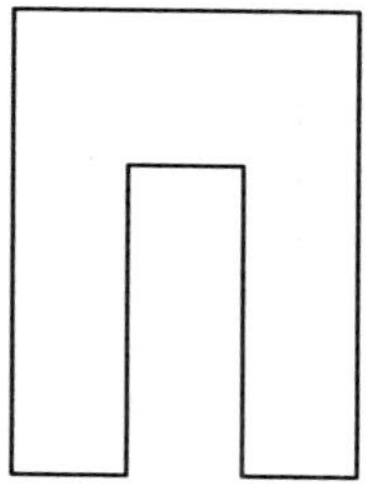   C 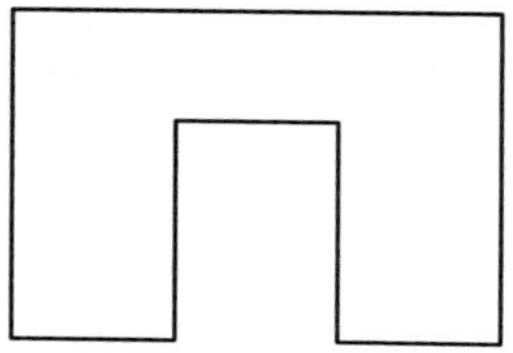   D

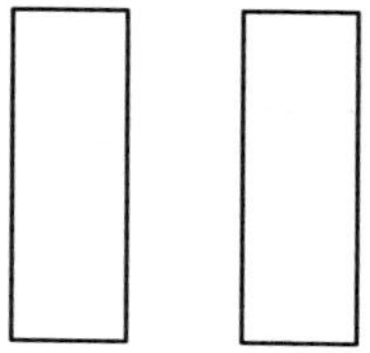

Name:

Class:

Section:

Date:

Developing Spatial Thinking

Grade:

Page
cp/cs-12

For the objects shown on the left below, circle the letters corresponding to all possible cross-sections that could be obtained by slicing it with an appropriate cutting plane. There could be more than one correct answer.

1.

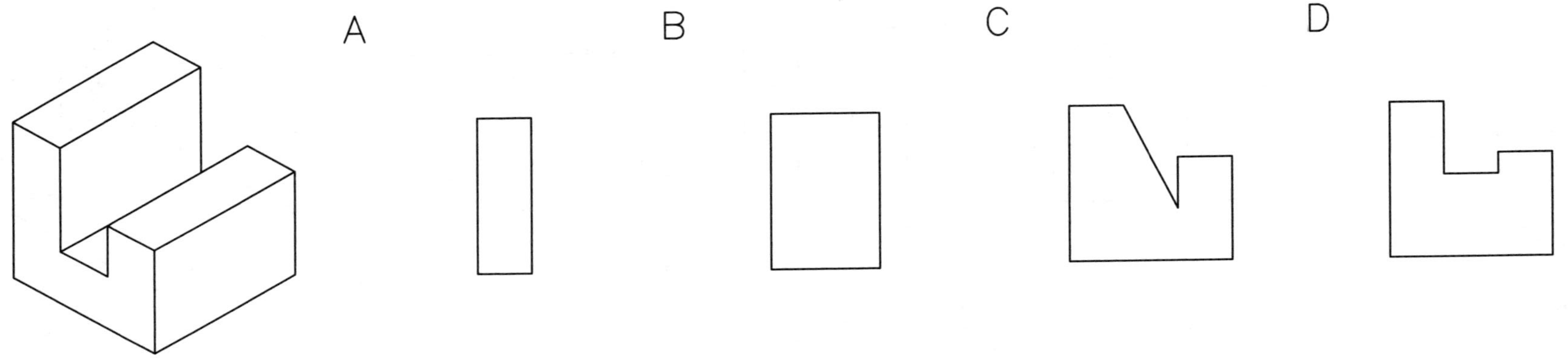

A   B   C   D

2.

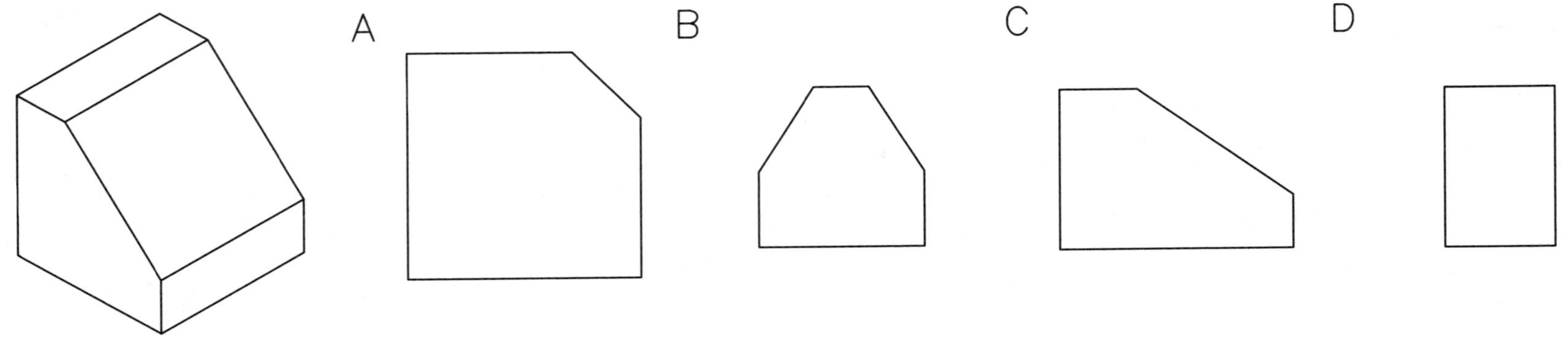

A   B   C   D

Developing Spatial Thinking

Page
cp/cs-13

Grade:

Name:
Class:

Section: Date:

For the objects shown on the left below, circle the letters corresponding to all possible cross–sections that could be obtained by slicing it with an appropriate cutting plane. There could be more than one correct answer.

1.

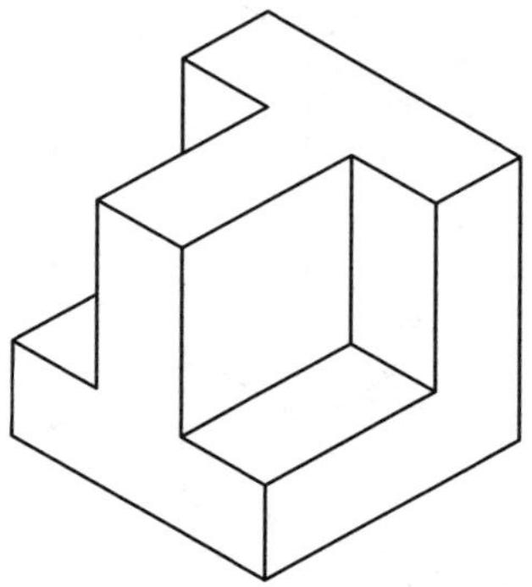

A

B

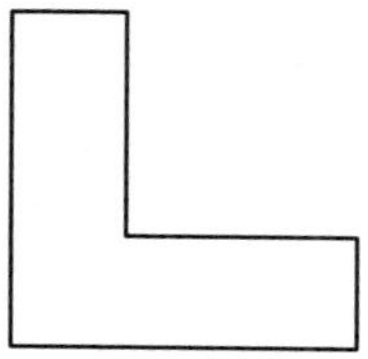

C

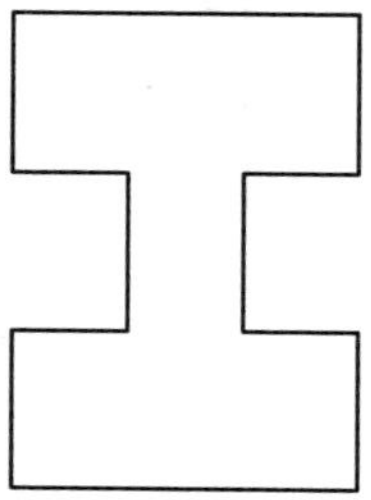

D

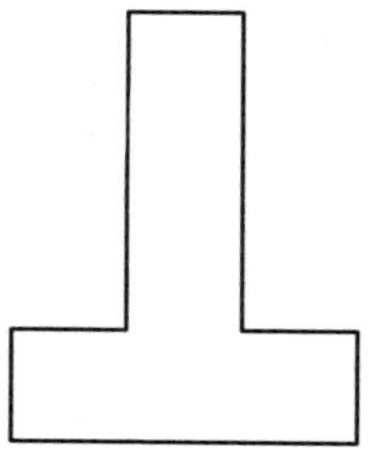

2.

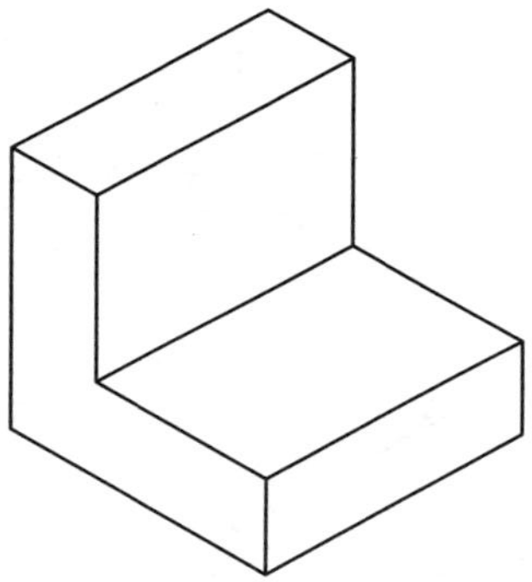

A

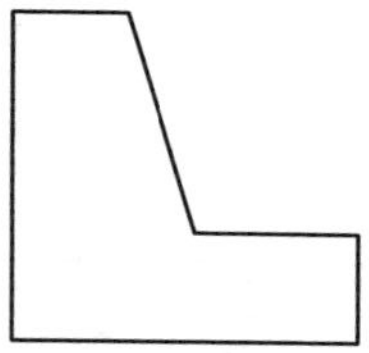

B

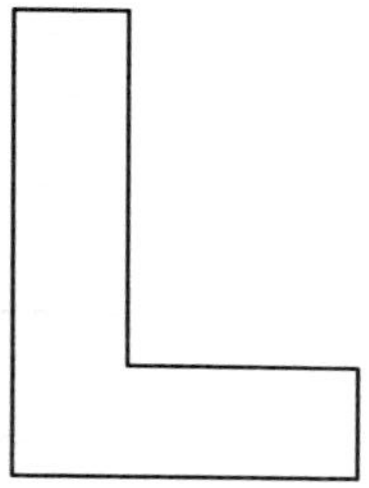

C

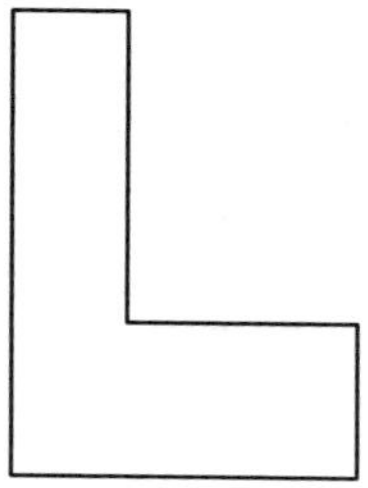

D

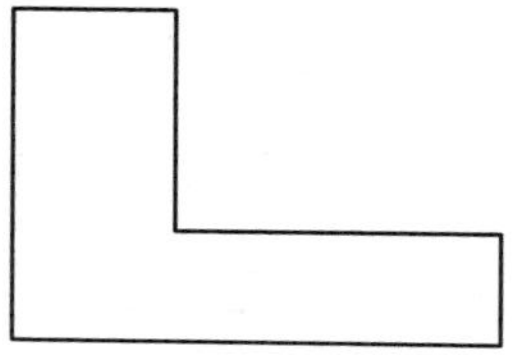

Name: Class: Section: Date:	Developing Spatial Thinking	Grade:	Page cp/cs–14

For the objects shown on the left below, circle the letters corresponding to all possible cross-sections that could be obtained by slicing it with an appropriate cutting plane. There could be more than one correct answer.

1.

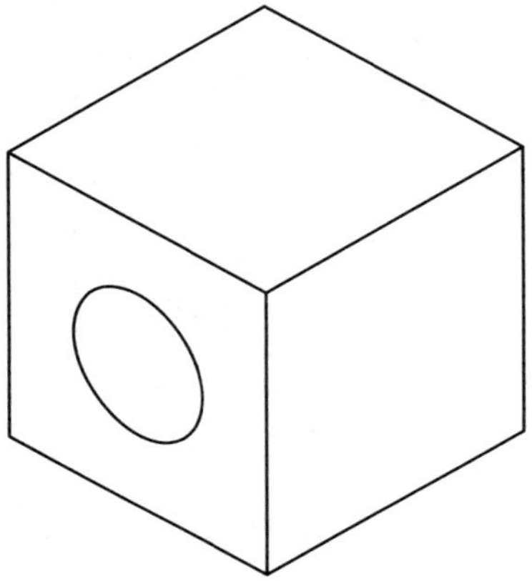

A 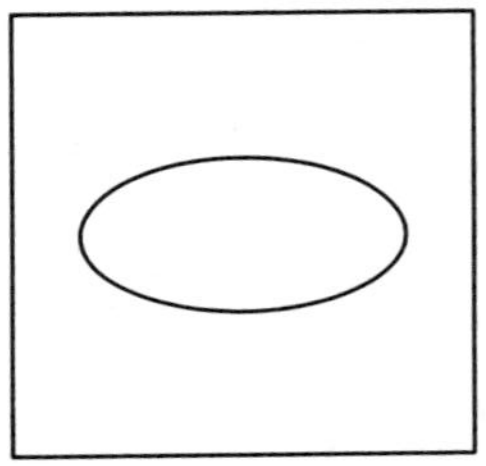   B 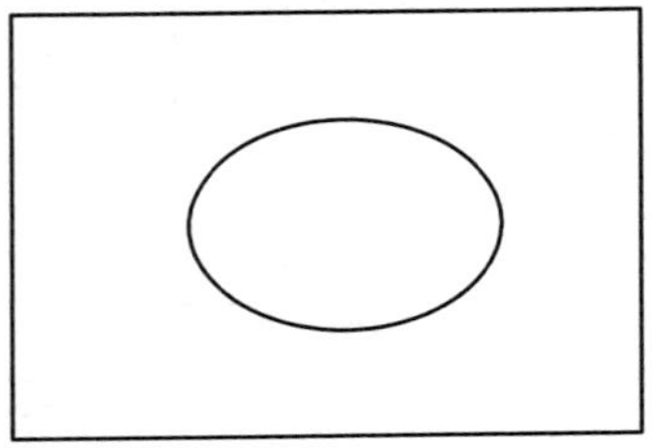   C 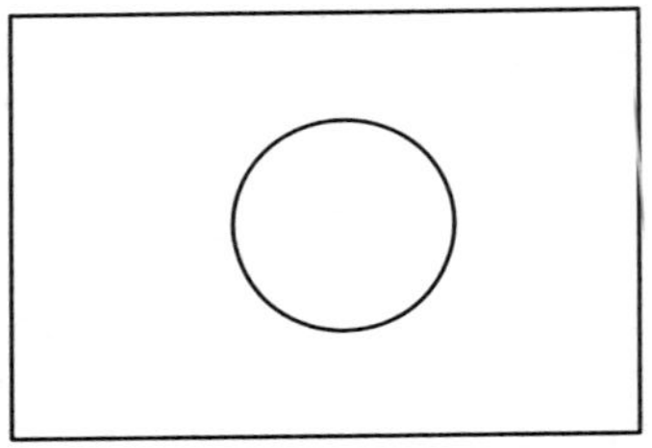   D

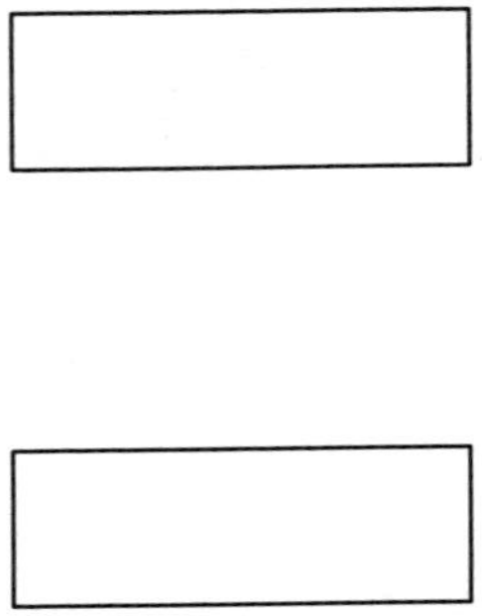

2.

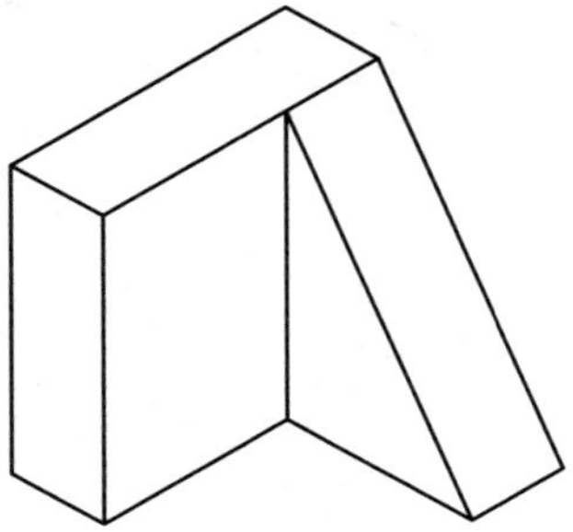

A 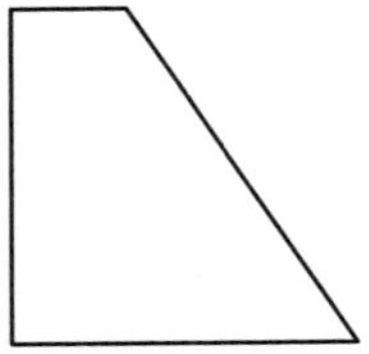   B   C 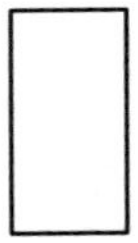   D

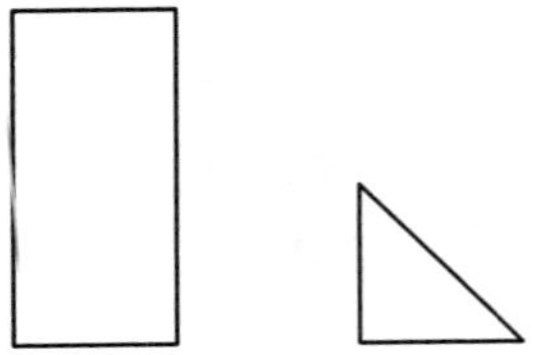

Grade:

Developing Spatial Thinking

Name:

Class:

Section:

Date:

For the objects shown on the left below, circle the letters corresponding to all possible cross-sections that could be obtained by slicing it with an appropriate cutting plane. There could be more than one correct answer.

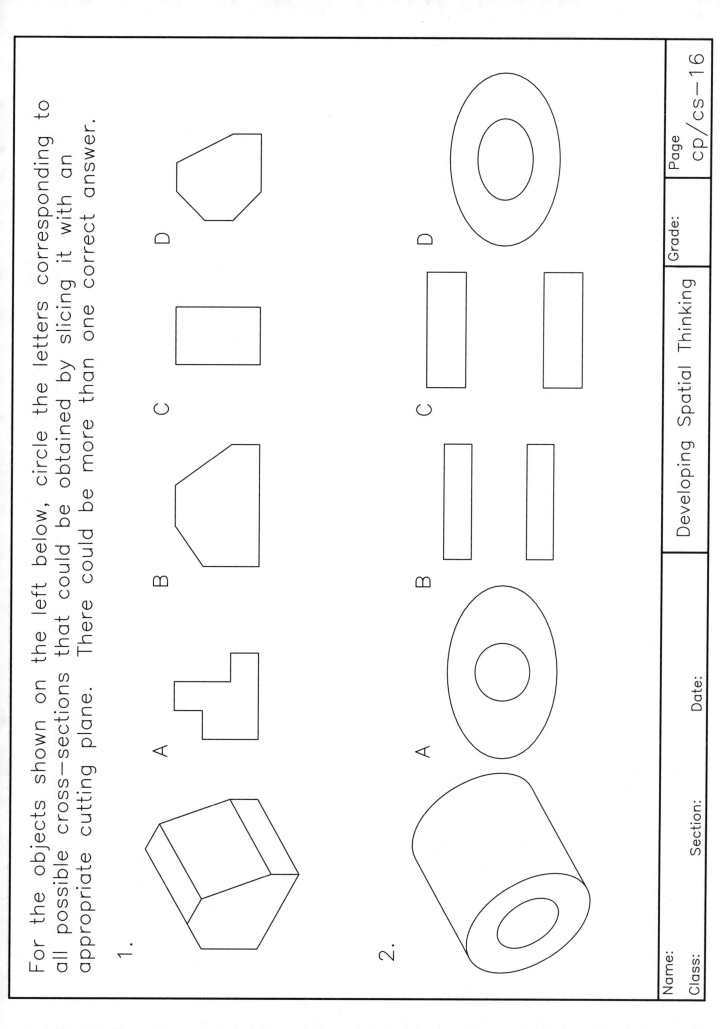

1.

A   B   C   D

2.

A   B   C   D

For the objects shown on the left below, circle the letters corresponding to all possible cross-sections that could be obtained by slicing it with an appropriate cutting plane. There could be more than one correct answer.

1.

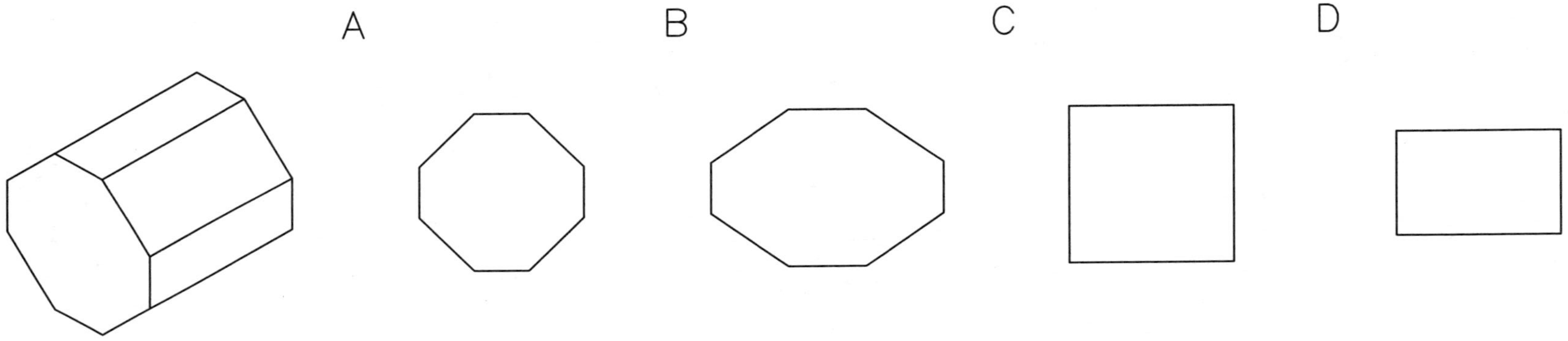

A   B   C   D

2.

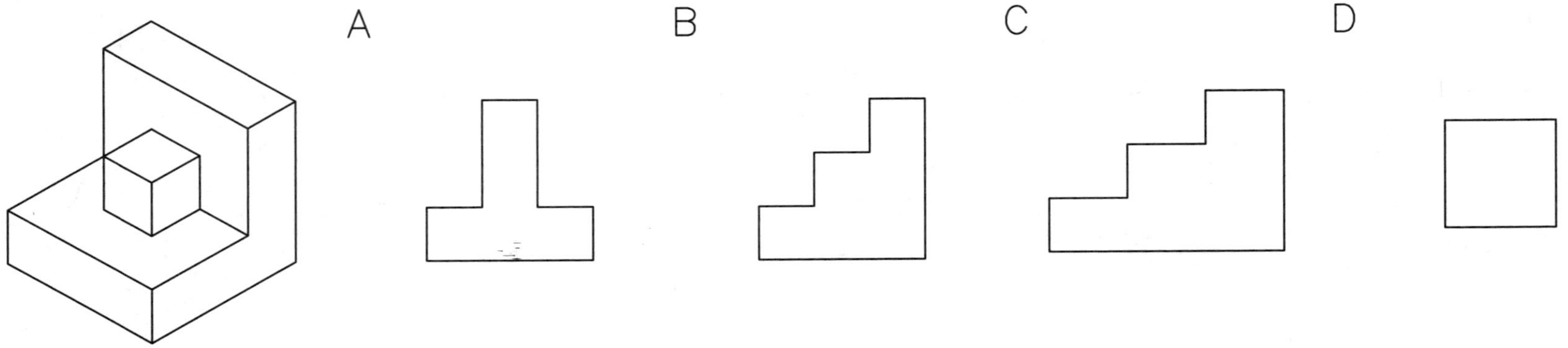

A   B   C   D

Developing Spatial Thinking

Page
cp/cs-17

Grade:

Name:

Class: Section: Date: